AF361360

BITTER HARVEST

BITTER HARVEST

An Inquiry into the War between Economy and Earth

Lisi Krall

SUNY
PRESS

Published by State University of New York Press, Albany

For information, contact State University of New York Press, Albany, NY
www.sunypress.edu

Library of Congress Cataloging-in-Publication Data

Name: Krall, Lisi, 1954– author.
Title: Bitter harvest : an inquiry into the war between economy and Earth /
  Lisi Krall.
Description: Albany : State University of New York Press, [2022] | Includes
  bibliographical references and index.
Identifiers: LCCN 2022005756 | ISBN 9781438489919 (hardcover : alk. paper) |
  ISBN 9781438489926 (ebook)
Subjects: LCSH: Environmental economics. | Human ecology—Economic aspects.
  | Nature—Effect of human beings on. | Agriculture—Economic aspects. |
  Agriculture—Environmental aspects. | Social evolution. | Capitalism.
Classification: LCC HC79.E5 K688 2022 | DDC 333.72—dc23/eng/20220505
LC record available at https://lccn.loc.gov/2022005756

10 9 8 7 6 5 4 3 2 1

For Florence Rose Bertagnolli Krall Shepard
and Kathryn Ann Krall Morton

Contents

Part III: Apogee

Preface

In 1996 the human ecologist Paul Shepard (my stepfather) died after a long battle with lung cancer.[1] In the week before his death when the line between what was real and imagined began to break down he turned to my mother one night and said that she shouldn't be alarmed if when she awoke he was not there—he would be in the backyard scything. At the time, it seemed to me a strange place for him to "wander." After all, he had dedicated his life to a critical and scathing appraisal of the impact of agriculture on humans and the Earth, and all that had been lost on both accounts when humans began to domesticate plants and animals. I can see now that in his last moments Paul toggled between two worlds. In the simplicity of scything, he could still imagine meeting the eye of a wolf at the edge of a field. He could still glimpse the "finely tuned" human ecology of our Pleistocene evolution with its demography of "a slow-breeding, large intelligent primate." Looking forward, he understood that a madness had overtaken us. An unimaginable future was unfolding, one unforeseen by those who scythed those first domesticated grains and ploughed the first fields.[2] Clearly agriculture was a point of departure from one way of being to another, and he was not finished thinking about it.

Little did I imagine, in that summer of 1996, that some twenty-five years later I would be engaged in deep contemplation over the agricultural revolution, its economic legacy, and the place of humans on Earth. As it turns out, this sort of contemplation has absorbed much of my academic life. It is probably accurate to say that Paul Shepard's influence churned up a skepticism about agriculture that resonated with me. That skepticism was further influenced by my contact with Wes Jackson.[3] It was Wes that said humans became "a species out of context" when they began the practice of grain agriculture. His incisive aphorism has always provided me with a guiding light. He, like Paul, recognized agriculture as "a fall," a point where the human relationship to the more-than-human world shifted in a problematic way. Wes calls agriculture a "fault line in human history," and he is correct about this. In fact, Wes says: "So destructive has the agricultural revolution been that, geologically speaking it surely stands as the most significant and explosive event to appear on the face of the earth changing the earth even faster than did the origin of life."[4]

I am by training a heterodox economist and inclined to think deeply about economic systems—what they are and how they come to be what they are. This interest intermingled with my acquired skepticism about the agricultural revolution to put me on the path of this inquiry. To better understand my orientation, simply juxtapose two distinct economic systems that will serve to highlight the profound differences in the human relationship to Earth that are framed therein. Hunting and gathering is an economic system where *Homo sapiens* lived as minimalists: surplus did not exist, feedback loops prevented expansion, and in material life humans were mostly independent and self-reliant (most could quite literally fend for themselves). Each human had an expansive knowledge of the more-than-human world (they were observant), and they used that knowledge to obtain their material necessities (food, shelter, clothing). One can argue it was an economic system embedded in the rhythm and dynamic of the more-than-human world and did not have feedback loops of expansion. Lest you disregard the importance of this system, please consider that humans lived in it in one form or another for two hundred to three hundred thousand years of their history as *Homo sapiens*. This is not irrelevant to our understanding of social and economic evolution.

Global capitalism can be described as a system where humans are not minimalists. In their productive life they are existentially interdependent (think about assembly line work, global supply chains, and global markets), and they are involved in a system dynamic that is expansionary, where surplus takes the form of profit and feeds an endless process of capital accumulation, exploitation, and crisis. This system structure and dynamic sets up a very real structural duality between humans and the more-than-human world. These two economic systems could not be more different, and they contextualize the human relationship to Earth in entirely different ways.

In the comparison of hunting and gathering and global capitalism it is evident that humans did not move directly from one to another. Looking simply from the perspective of population dynamics and the incidence of ecological collapse it is evident that the cultivation of annual grains (the agricultural revolution) marks an inflection point between the two. The agricultural revolution around annual grains recalibrated human economic order and the human relationship to Earth. Surplus and expansion, hierarchy, profound material interdependence around the focal point of grain production, and powerful feedback loops between population, grain (energy) production, and division of labor helped to form a distinct economic system. The cultivation of annual grains was not simply a change in the way humans secured food: it was an entirely different economic system.

Like many others, I began to contemplate deeply the etiology of this monumental economic system change. This is where a simple observation came to bear on my thinking and influence my approach: humans are not the only species to engage in agriculture, and for those species that do, the structure and dynamic of their agricultural systems, which is to say their economic systems, look very similar. I wondered whether looking more closely at other species might be of benefit in understanding economic formation in humans and in particular this profound shift to agriculture. As you might imagine, this took me down the road of a deeply materialist and evolutionary approach to understanding the formation of the agricultural system or what I eventually labeled the *economic superorganism*.

Had it not been for the intellectual companionship of John Malcolm Gowdy it is likely I would not have walked down this road. John and I engaged in foundational research together. This initial research provided the springboard to this book. As a progressive social scientist, I found it difficult to seriously consider that economic systems might be influenced by processes of evolution. The deterministic undertones this approach suggests were uncomfortable, especially for someone who believes system change is essential. In the end, I took my inquiry out of disciplinary boundaries and came to a new understanding of economic systems and the present war between economy and Earth.

In commenting on the transdisciplinary nature of his book *The Columbian Exchange*, the late environmental historian Alfred Crosby recognized that "historians, geologists, anthropologists, zoologists, botanists and demographers" would see him "as an amateur in their particular fields."[5] My list would not be precisely the same as Crosby's, but I am sure in what follows the experts in the fields I navigate might level the same complaint at me. I certainly challenge the boundaries of traditional historical analysis and evolutionary biology in addition to economics in the pages that follow. My purpose is not to engage disciplinary debates but instead to "drive knowledge out of its categories" as my friend Wes Jackson would say. In doing so I want to create an opening for rethinking and reimagining the human place on Earth, and for a more expansive understanding of the challenge of altering the economic trajectory in play. My hope is to replace hubris with humility, the preoccupation with scarcity with a preoccupation for limits, and the focus on sustainability with a focus on real rapprochement with Earth.

Acknowledgments

I am indebted to my mother, Florence Shepard, and my sister Kathryn Morton for the years of conversation that helped to hone the ideas that undergird this book. Their support has been indispensable. I am grateful to John M. Gowdy with whom I engaged in foundational research on ultrasociality. That work provided a springboard into *Bitter Harvest*. My contact with David Sloan Wilson and the Evolution Institute facilitated the formation of my own thinking on evolution and economic systems. I have benefitted from the ongoing conversation with the many people who have been drawn to revolutionary and transdisciplinary thinking and a love of Earth. Most especially I acknowledge the influence of those associated with the Land Institute, Ecosphere Studies, and the New Perennials Project. That bunch includes but is not limited to Wes Jackson, Joan Jackson, Bob Jensen, Bill Vitek, Stan Cox, Priti Gulati Cox, Aubrey Streit-Krug, Tim Crews, Eileen Crist, Marc Lapin, Nadine Canter Barnicle, and Michael Johnson. The careful and incisive comments provided by Stan Cox, Eileen Crist, and Bob Jensen were especially beneficial in the final drafting of the manuscript. The New Perennials Project and the Surette Fund provided support that came at a critical juncture in the project.

I have been blessed with many good colleagues/friends that have provided support and guidance, in addition to venues for vetting my ideas. I include among them Kevin Mayer, Sharon Steadman, Girish Bhat, Scott Moranda, Jaroslava Prihodova, Jamie Dangler, Ann Vittoria, Julia Ganson, Flavia Dantas, Jessica Carrick-Hagenbarth, Avanti Mukherjee, Katherine Graham, Brian Ward, Kathleen Burke, Kent Klitgaard, Victoria Boynton, Mark Prus, Jean Reagan, Julia Corbett, and Josh Farley.

Finally, I am indebted to my son, Francis Prus, who has always managed to impart wisdom and a sounding board for discussing the difficulties entailed in the creative process and its path.

Introduction

But life involves before everything else eating and drinking, a habitation, clothing and many other things. The first historical act is thus the production of the means to satisfy these needs, the production of material life itself . . . This connection is ever taking on new forms, and thus present a "history" independently of the existence of any political or religious nonsense which would especially hold men together.

—Karl Marx, "The German Ideology, Part I"

How did *Homo sapiens* come to this moment in their history where the order of economic life is at war with the Earth? The more-than-human-world has been eclipsed and degraded by an unassailable global economic system run amok. Climate change, the sixth great mass extinction, soil erosion, depleted groundwater, and toxicity are but a few examples of a seemingly irreparable turn. We know what is happening as we endlessly document this downward trend. Yet despite this knowledge we have not altered this trajectory, leading one to conclude that it is the structure, momentum, and power of the economic system that governs the forward march. We appear to be captured by a system largely impervious to attempts to alter its course.

This is the tragic economic history of *Homo sapiens*, and yet there is ample historical evidence of humans providing themselves with food, shelter, clothing, love, technology, and art for many tens of thousands of millennia in greater compatibility with the more-than-human world: quite literally living as one of many. The central focus of this inquiry is to explore a broad arc of economic history with an eye to this change; that is, to the duality that has been created between humans and Earth. There has been a decisive turn in human history where humans went from being a species embedded in the rhythm and dynamic of the Earth in their economic life to the opposite. That decisive turn was the agricultural revolution around annual grains.[1] It is the point where a distinctly different economic system took hold, creating a duality between humans and Earth. Here *Homo sapiens* began an evolutionary experiment as an economic superorganism. Mostly unrecognized in its importance among scholars of our present ecological/economic crisis is this fundamental shift in economic order (collective material life). What came after is derivative.[2]

The narrative on agriculture has been mostly caught up in a "just-so-humancentric" story. In one form or another it usually goes like this. Smart human beings, in the endless quest for survival, invented agriculture. Some go on to claim that agriculture was but another step on the road to progress. It brought *Homo sapiens* out of the cave and into the light of civilization. The approach taken here runs against this current. I see agriculture as a universal system engaged by many species. In the shadow of the universal I move away from a just so human story toward a more expansive understanding of the etiology and importance of the profound change in human economic organization that took hold with agriculture. Humans were *collectively* reconfigured and their relationship to Earth deracinated with agriculture. The legacy of that change now has come to rest in global capitalism. Thus, the present collision course between global capitalism and Earth should be interpreted as a particular manifestation of a system change that has been in motion for ten thousand years.

Most conversations about the economy and the ecological crisis concentrate on some combination of the triumvirate of the industrial revolution, technology, and capitalism as the confluence of forces responsible for our problematic path. Scholars of our present predicament have been captured by the power of the dramatic exponential flight that followed this confluence, sidelining clues to the present that reside in the past. The importance of the agricultural revolution all but recedes into the realm of obscurity, never fully discounted but never wholly acknowledged either. So too any relevance the long span of human history that predates it might have in our thinking about the present.[3] The seed of the human expropriation and domination of Earth goes much deeper than seams of coal or the recent arrangement of economic life known as capitalism. A more foundational and incisive understanding of the phylogeny and etiology of our present economic system and the challenge it presents in our relationship to Earth is found in the past and the connection to the agricultural system.

Grain agriculture ushered in many aspects of modern economic life that have only become more pronounced in the ensuing ten thousand years: surplus and expansion, ecological decay, inequality and hierarchy, extreme material interdependence, as well as a structural duality between humans and the more-than-human world. We are wholly dependent on Earth, and yet we are no longer in community with it. We are profoundly interdependent, and yet we are no longer in relationship. We are engaged in an economic system that is essentially self-referential on an Earth that is relational. Thus, we now reside in contradiction and paradox. This center will not hold. The frenzied pitch of discordant realities is not sustainable, and we are now approaching a fateful divide where we are rapidly eliminating other species and the impulse of all that is not us.

A few methodological questions must be explicated in order to understand the logic and execution of this book. I begin with a detour down the road of evolution (or what I consider a deeply materialist approach to understanding the formation of the agricultural system). Some might be thrown off by this starting point, so let me reiterate the purpose. There are many species that cultivate, and the structure and dynamic of the economic systems that develop out of cultivation are strikingly similar. I chose not to leave hanging these "universalities" surrounding agriculture. And in engaging what is universal I tapped into evolution simply because it seemed to be a logical place to look for explanation and etiology. This path took me into debates in evolutionary biology surrounding the evolution of sociality and cooperation. In the end I raise questions about human cooperation, its place in economic systems, and the place of economic systems in the matrix of evolution that only a novice would raise. My intent is not to sort out debates in evolutionary biology but to create a plausible story about the emergence of a universal system, a new collective "whole" that altered the human relationship to each other and to Earth.

I also engage history in this inquiry, and a few words must be said about how I go about that. My historical approach is likely to make historians uncomfortable. They are scholars of detail. As an economist, some abstraction from the detailed meandering of history in order to find the patterns of economic life is a necessary tool of my trade, and I take full liberty with that inclination in the pages that follow. I am interested in understanding the emergence of an economic structure and dynamic that created a duality between humans and Earth, and in this I weave the tapestry and logic of a long arc. I move away from the detailed meandering of history to elucidate an altered pattern and force to economic life that took form around grain agriculture, and then I follow that pattern to the present.

A preliminary map of the story/contemplation that unfolds in the pages that follow will help orient the reader. I offer a deeply rooted economic focal point to understanding how we came to this war with the Earth and what appears to be an inability to do anything about it. Bitter Harvest is organized into four parts: "The Economic Superorganism," "Bitter Harvest," "Apogee," and "Epilogue." I reiterate—my purpose is to highlight the story of the emergence and power of an economic system that created a duality between humans and Earth that had not previously existed and is with us now in exaggerated form. I label this system the *agricultural* system and also an economic superorganism to denote its power. I highlight the forces that led to this inexorable change as well as its significance. An economic system contextualizes the human relationship to the more-than-human world, and that context was foundationally altered beginning with grain agriculture. This inquiry opens the door to humility rather than hubris in our approach to our problematic relationship with Earth, and it connects rather than separates humans from other species. Uncomfortable questions about determinism and the power of economic systems emerge in these pages. My intent is not to discourage action but to more expansively frame the challenge we face.

The first section of the book (Part I), "The Economic Superorganism," begins with a simple observation: humans are not the only species to cultivate. Interdependence forged through a division of labor around the focal point of energy production, expansion, and the emergence of particularly powerful feedback loops are common system characteristics, indeed universal characteristics, found in diverse agricultural species. In concert they form a powerful universal system. I utilize a transdisciplinary methodology where I delve into the evolution of cooperation to parse the building of the universal system of agriculture, which is a complex matter involving synergies of collective evolution. I engage the possibility that the exceptionality of humans and all of the "just so" stories we tell ourselves about the human transition to agriculture and the ordering of economic life might need to be reexamined; I also raise the disturbing question of determinism in the formation of the economic superorganism.

Part I enters into rather esoteric debates in evolutionary biology not as an academic exercise but because it is impossible to carry out the nuanced discussion entertained here without the benefit of some investment in the technicalities of evolutionary theory particularly as it pertains to the formation of groups and the evolution of sociality and cooperation. This section raises two important questions pertinent to evolutionary theory: What role does evolution play in the formation of economic systems? Where do those economic systems lie in the matrix of evolution? I am led in this discussion to ask whether humans are crossing an evolutionary threshold as they engage in the mass extermination—the sixth extinction—of other species on the planet and human population and the manifestations of human material life overwhelm Earth. Part I consists of three chapters.

Chapter 1, "Agriculture and the Evolution of the Economic Superorganism," taps into evolutionary theory to understand the formation of the agricultural system in light of the fact that agriculture is not the exclusive domain of humans. Insect and human agriculturalists (species that could not be more different on an individual level) were collectively reconfigured in a very similar way around agriculture and through similar evolutionary processes. With this in mind chapter 1 engages the literature surrounding the evolution of cooperation and uses this approach to bring the light of the universal to our understanding of the agricultural system—what I have also labeled the "economic superorganism."

All species that engage an agricultural system become cooperative in a universal way: a structure and dynamic to cooperation emerge through a division of labor that centers on the focal point of cultivation (energy production). Powerful feedback loops develop between division of labor, population, and energy production. This chapter leads us to ask whether this collective configuration is rightly viewed as a powerful whole in the matrix of evolution. It is also clear that the agricultural system moved humans inexorably away from a fluid interchange with the rhythm and dynamic of the more-than-human world in the day-to-day provisioning of economic life. Their material life is refocused within an emergent solipsistic collective system. A duality between humans and Earth takes form as well as a foundational change in the expression of human cooperation.

Chapter 2, "Agri-culture?," expands the discussion of human cooperation as it pertains to the formation of an agricultural system. This chapter makes the case that it is not accurate (in the case of humans) to see the formation of the agricultural system and its ultra-cooperation simply as a vestige of the evolution of the human capacity for culture. Cooperation, as it takes form in the agricultural system through the division of labor, is not adequately captured by culture. The division of labor is characteristic of species life and not the exclusive domain of humans, and it is universally extended in agricultural species. It plays a central role in the formation of the economic superorganism creating a profound interdependence around a central focal point of food production (cultivation) regardless of the species. Humans are unique in their capacity for culture, but it is possible that culture has created opacity in our understanding of more universal aspects of cooperation that form through a division of labor. This chapter challenges the belief that economic order is merely a vestige of culture, and it raises the possibility of a determinism in economic life that is intended to disturb the reader.

Chapter 3, "The Division of Labor," enters the intersection of evolutionary processes and economic formation in more detail in order to demonstrate that the engagement of agriculture involved the coevolution of all species involved. The emphasis here is on the way the humans and insect cultivators were changed in the process. Cooperation forged through a division of labor around the focal point of cultivation reconfigured the agricultural group helping to make it an integrated whole. The division of labor around the focal point of food production is viewed as an emergent characteristic of the agricultural system—it provides a core structural stanchion to the formation of the economic superorganism. Here culture serves as a mechanism to engage the division of labor in humans just as mutation and selection are the mechanisms used by insect agriculturalists.

The second section of the book (Part II), "Bitter Harvest," navigates the distance from the universal to the specific, more fully accounting for the particularities of the human transition to agriculture. This section of the book integrates the complexity of evolutionary processes, culture and ingenuity, chance circumstances, and the power of the universal system into whole cloth; that is, into an integrated economic system (an economic superorganism) that began with grain agriculture. Humans became a self-referential and profoundly interdependent species of expansion and surplus with agriculture—just as their insect counterparts had—but did so with their own imprint and their own history. The particular coevolutionary dynamic between annual grains and humans and the human propensity for institutional life is elaborated and integrated with the universal characteristics of the economic superorganism (division of labor, population growth, and energy production) to create the particular tapestry of the human economic superorganism. The reticulated and self-reinforcing nature of grain agriculture as an integrated economic system becomes clearer as does the duality it establishes between humans and Earth.

Part II offers a reinterpretation of agriculture, not as an inevitable trajectory toward civilization and progress but as a problematic turning point in the evolutionary history of humans. I recognize that agriculture brought humans civilization—a benefit mostly to the few that flourished in the many advantages of being on the receiving end of its surplus. My homage is to the majority of humans who were enslaved directly and indirectly in relentless sweat, toil, and alienation in their daily lives through the agricultural system; to the Earth disrupted, interrupted, and temporarily diminished in its established cycles and complex ecologies; and to the foundational and sacred connection between humans and the more-than-human world that was undermined by the emergence of the economic superorganism. The transition to agriculture created an economic system where humans were no longer in community with Earth. Part II consists of two chapters.

Chapter 4, "The Tapestry of the Universal and the Particular" weaves a tapestry of the universal and the particular in the human transition to grain agriculture. It becomes clear in this chapter how the coevolution of humans and annual grains gave rise to the universal system and how that system was extended by unique human attributes. An extensive exploration of coevolution is undertaken in this chapter. Chapter 4 also introduces the institutional and cultural trappings of surplus (hierarchy, patriarchy, slavery, markets, debt, money, taxes), and the human capacity for inventiveness and explores the way they extend the expansionary and self-referential tendencies of the agricultural system. Among the institutional elaborations of the agricultural system the economic institutions stand out (markets, expanding trading networks, debt, money, taxes, property rights). And these institutional embellishments take on a life of their own.

Chapter 5, "A Species Out of Context" is a phrase borrowed from Wes Jackson to describe agriculture as "the fall." This chapter explicates the fall through an elaboration of its effect on humans (both individually and collectively) and their relationship to Earth. The agricultural system altered the expression of human life—the relationship to the more-than-human world—and it changed the material dynamic of humans on Earth. The inclinations of the agricultural system are brought into focus: surplus, interdependence, duality, collapse.

Hunting and gathering as an economic system is juxtaposed with the agricultural system to highlight the contextual nature of the human relationship to the more-than-human world and the profound change that agriculture entailed. With agriculture humans were no longer the nonexpansionary, minimalist species embedded in the rhythm and dynamic of the more-than-human world that they had been for the vast sweep of their history. Instead humans had become an expansionary, accumulative species, embroiled in an economic system with powerful feedback loops, making the system a thing unto itself and inclining it to ecological overshoot. A dramatic change in the expression of human life on Earth took place with the transition to agriculture that was not centered in the human genome; yet one can argue that *Homo sapiens* became *Homo sapiens agriculturii*, members of the particular human economic superorganism that enslaved many, thwarted human expression, all but undermined individual autonomy, and established a structural duality between humans and Earth that had not previously existed.

The third section of the book (Part III), "Apogee," highlights the final and dramatic expression of the economic superorganism that takes form in capitalism. It is a whole system with its own integrity, but it is also the legacy of the agricultural system. In this sense it is appropriate to see capitalism as a system within a system. Capitalism changed the form of surplus and expansion but not the fact of their existence; it altered human-to-human relationships in material life but did not change the fact of enhanced material interdependence (or the presence of hierarchy). Finally, capitalism drove the wedge of duality between humans and the more-than-human world ever deeper, but it did not create that duality.

Yet capitalism is its own whole, and once it is fertilized with the industrial revolution, the duality between humans and Earth that began with agriculture takes on a most pernicious form. An expansionary and self-referential system is now freed from energy constraints, and the potential for crises embodied in its institutional structure becomes ever more formidable as the system matures around fossil fuel, as does the potential for ecological overshoot. The paradox presented by the economic system is formidable; growth is required for jobs and fighting stagnation, and degrowth is required for staying within biophysical limits. The profound duality of capitalism resides not only in the economic system but is reflected in the economic thinking of the past 250 years. Part III consists of two chapters.

Chapter 6, "Capitalism: A System within a System," provides a detailed discussion of the formation of capitalism and helps the reader understand the particular ways that human capacities for institutional life and inventiveness elaborated surplus and intermingled with universal processes and forces to form this particular variant of the agricultural system. The combination of capitalism (a particular institutional embellishment of surplus) and the industrial revolution (a technological innovation that created a seemingly infinite supply of energy) bring the trajectory of the economic superorganism to its apogee. Capitalism is both a supra-material system (disconnected from Earth), and a material system (profoundly connected to Earth). This is the paradox and duality we now confront. An economic system is always material, but capitalism functions as a supra-material system where economic variables interact as if disconnected from the Earth. Internal crises—that is to say crises internal to the system (an interruption of the circular flow of income and spending, for example), dominate the economic landscape.

Chapter 7, "In Search of a Deep Ecology of Economic Order," critically assesses economic thought in light of the rise of the economic superorganism and its present form. Over the past 250 years the ideas of the great economists (the "unearthly philosophers" I call them) have orbited the supra-material aspect of the economy, and any connection to the Earth has existed at the margins of their analyses. Ironically a discipline erected to understand material life is removed from an analysis of the material dimensions of economic life. In time a new group of economic thinkers, those intent on reconnecting the economy to Earth, has entered economic discourse (I refer to them as the "earthly" philosophers). Yet even among this group the importance of fully appreciating the long arc of history has remained elusive, and so too a clear focus on the challenge and complexity of the twilight of the legacy of the agricultural system. The challenge is to move to a deeply ecological economic system where humans take their place as one of many species that inhabit Earth and to engage in more humility and reflection when approaching the economic superorganism.

The final section of the book, "Epilogue," consists of only one chapter. "Languishing in the Twilight of the Apogee" offers a final reflection on the odd evolutionary history of humans where they find themselves at once an economic superorganism and a Pleistocene species. Humans are now caught up in contradiction and stand at a divide in their evolutionary history and also with regard to the fate of Earth, its wild impulse, and its self-willed otherness. History tells us that our levers of change (culture, institutional life, inventiveness) generally work with a system, not against it. It is therefore essential to focus on what we intend with them, especially in the twilight of the apogee of the economic superorganism. The cumulative nature of the past ten thousand years is upon us, and in this we are simply forced to face the prospect of slipping down the other side of the divide, taking an irrevocable turn as *Homo sapiens agriculturii* and finishing the sixth mass extinction.

I add here a "Glossary of Terms" for easier navigation of *Bitter Harvest*.

Glossary of Terms

Agriculture: In this book agriculture refers specifically to annual grain cultivation in the case of humans and fungi cultivation in the case of insects.

Autocatalytic: This is a term borrowed from chemistry. Here it simply means that the feedback variables of the agricultural system (population increase, cultivation, division of labor) react with one another in an expansionary dynamic.

Capitalism: The contemporary and institutionalized form of the economic superorganism.

Duality: Duality is used in this book to refer primarily to the rise of the economic superorganism in humans and to place that system in reference to its relationship with the Earth. When duality develops, the system tends toward ecological breakdown as humans are no longer embedded in economic life in the rhythm and dynamic of the more-than-human world. Duality becomes so exaggerated with capitalism that it is a system that is simultaneously two things—a supra-material system functioning in a self-referential way apart from Earth and, at the same time, a material system with profound demands and impacts on the ecologies of Earth.

Earthly philosophers: Those economic thinkers that focus on reconnecting the economic system to its biophysical foundations.

Economic superorganism: The economic system put in motion with the practice of grain agriculture in humans and fungi production with certain insect species. The system is a configuration of powerful feedback loops especially between cultivation, population, and division of labor. These feedback loops enhance and reinforce each other, creating interdependence and expansion. The agricultural system is an economic superorganism but later this system itself evolves in the case of humans to take the form of global capitalism.

Homo sapiens agriculturii: This is my terminology, as I know of no one else that uses it. It is meant as a reference to humans who become members of an economic superorganism in order to reinforce the idea that humans became something distinctively different when they became organized around the agricultural system. It is possible the transition to agriculture was a major evolutionary transition for the species that engaged in this mode of production.

Paradox: When I refer to paradox I am referring specifically to the circumstances of the present where the duality of the economic system is so pronounced that the system requires growth for jobs, and degrowth for containment within the ecological boundaries of Earth. There is no easy resolution to this situation.

Self-referential: The use of this term is simply meant to convey that the agricultural system functions as a system. It is another way of saying an economic superorganism forms around agriculture or that agriculture involves an autocatalytic dynamic. Referring to the agricultural system as "self-referential" is meant to reinforce the idea that it is an insular system with feedback loops particular to it.

Unearthly philosophers: Those economic thinkers who attempt to describe the economic system in its supra-material form: that is, in the way it functions as a system removed from its material roots.

Part I

The Economic Superorganism

Chapter One

Agriculture and the
Evolution of the Economic Superorganism

The human transition to agriculture was a major economic transition in human history. The transition to the cultivation of annual grains changed the nature of human material life so fundamentally that we can think of it as a fault line in human history and the beginning of a trajectory that led to our present moment—to this war between the economy and Earth.[1] It was a categorically different expression of collective species life (of economic life) and not simply a gradual change along a continuum where humans change the environment, and the environment changes humans over time. Instead it was an entirely different path.[2] Grain agriculture changed the expression of human cooperation, the collective energetics of the human species, and the human relationship with the more-than-human world. It formed an economic system that had an insular integrity.

Until the agricultural revolution humans had been a Pleistocene species; minimalistic, nonexpansive, individually self-reliant, observant, and embedded in the rhythm and dynamic of the more-than-human world. Humans were but one of many species on Earth. Humans became materially—and therefore existentially and structurally—interdependent and expansionary: a duality between them and the more-than-human world began to take form with the cultivation of annual grains. In rough outline all of these characteristics of economic life remain with us. One can argue that we are now confronted with the final crescendo of the legacy of the agricultural revolution in global capitalism and a human population of almost eight billion people where mass extinction and climate change are the order of the day. It's difficult to reconcile these facts with the claim that humans have been on a trajectory of progress.[3]

The fundamental change in human collective life expressed in grain agriculture stands before us demanding explanation, just as the evolution of cells, organs, bodies, and the rest of the agglomerations of life that come into existence and have integrity and force in the unfolding story of life on Earth demand explanation. The reason I am inclined to engage evolution to understand the structure and dynamic of agriculture and the full story of the human transition to it owes to a simple observation: humans are not the only species to have made the transition to agriculture. In other words, what appears at first glance to be uniquely human is not. More importantly, all species that practice agriculture express a remarkably similar structure and dynamic in their collective material life around this mode of production.

In my exploration of agriculture, I am nudged to adopt a deference and openness to what may or may not be the exclusive domain of humans and in doing so to think more expansively about the processes and mechanisms giving rise to human economic order and, more specifically, to the profound reordering of economic life that occurred around grain agriculture. Let me begin by simply describing the universal characteristics in the collective life of all species involved in agriculture. All engage in the collective production of food through cultivation. All engage a coevolution that plays on and enhances the inherent tendencies of the species that cultivate—this is especially apparent in the division of labor that becomes more elaborate and interdependent around the focal point of cultivation. All engage formidable feedback loops generated *within* the agricultural system. These feedback loops create a dynamic interplay involving food production, the division of labor, and population growth, making species that cultivate expansive and, for want of a better term, self-referential. They all become a materially integrated whole where individual autonomy in material life is dramatically diminished, if not eliminated altogether. Individuals have no choice but to engage in the collective agricultural enterprise and play their role in the system. If they don't, they won't survive. These are attributes of agriculture, no matter the species. They reveal universal inclinations in economic organization that crystalize with agriculture, thus making agriculture a universal system.[4] These are clearly evident in human societies that form in a relatively short period (by evolutionary standards) after humans began the cultivation of annual grains.

It is absolutely true that humans differ from the many other species that practice agriculture. Humans engage culture and institutional embellishments (e.g., hierarchy, markets) around agriculture, and they erect invasive technologies (e.g., irrigation, plowshares) that enhanced the agricultural economic dynamic giving rise to a more complex and powerful human experiment.[5] Humans make misery for many in the wake of the surplus created around grain agriculture. Hierarchy, patriarchy, and slavery are the most obvious examples. In the end humans have the capacity to reflect on themselves, and they experience a struggle with the tensions created by the agricultural economic order and its legacy, and other species do not. I would say we struggle with this now as we try to figure out how to become an ecological species when we are clearly caught up in a system that has forged a problematic duality between humans and Earth.

Despite these unique human attributes there remains the fact that there is a structural and dynamic integrity to agricultural societies that is common to all the species that practice it. The cross-species exploration did not take me to our closest relatives, the primates, who do not practice agriculture. Instead, it took me to social insects, specifically many species of ants and termites, that do. These species have an aptitude for collective material life, developed around the cultivation of fungi, that easily rivals that of humans after humans engaged grain agriculture. Among the people who are aware of this fact, there are few who have bothered to draw its relevance to the economic life of humans because, after all, humans are so different from these insect species.[6] There is no question these insect species and humans could not be more different on an individual basis; however, in the structure and dynamic of their material life around cultivation, the similarities are nothing short of astounding. In fact, the only reason we ignore this similarity is precisely because the individual differences are so profound. Clearly, it is not the comparison of the individual members of these diverse species that demands a closer look; instead, it is their ability to *collectively* engage an agricultural system. Grain agriculture, like fungi production in insects, is not something understood as an individual matter: it is a collective enterprise. So, when I look to evolution for guidance in understanding the universality of agriculture and its relevance to human economic order it is the evolution of the collective that is of relevance to me.

The similarity between insect superorganisms, especially those that practice agriculture, and human societies after they began the practice of grain agriculture has not been entirely ignored. Nor has their common remarkable sociality for that matter. For example, in the review of E. O. Wilson and Bert Hölldobler's book on insect superorganisms, Tim Flannery states: "Indeed it is the changes wrought in attine societies by agriculture that the principle interest for the student of human societies lies."[7] Hölldobler and Wilson are two of the most preeminent entomologists/evolutionary biologists that study the evolution of insect superorganisms.[8] E. O. Wilson later wrote the book *The Social Conquest of the Earth* in which he places both insect superorganisms and humans in a similar category of highly social and cooperative animals that have been extremely successful; they have literally taken over the world.[9] Unfortunately the common expression of sociality of insect superorganisms and humans around agriculture has never been parsed seriously either by evolutionary biologists or economists.[10]

Agriculture embodies a unique expression of cooperative behavior in the material lives of the species that practice it. Wilson and Hölldobler are very clear that insect agriculturalists develop the most complex societies of the biological superorganisms, so they do acknowledge agriculture in this sense. Yet as a matter of evolutionary significance agriculture has not been considered distinctive either for insects or humans. It is a sidebar to the history of their respective *social evolutions*.

I will come back to a detailed description of the extensive cooperation displayed in agricultural insects. But first it is essential to understand how evolution deals with the complexity of cooperation. A minimal investment in this literature is essential to this discussion. I begin with some clarifications about definitions and categories used by evolutionary biologists to discuss sociality and, by extension, extreme cooperation. E. O. Wilson tells us that both insects and humans are "technically comparable" as eusocial (truly social) species. Eusociality is defined by Wilson as sociality where groups contain multiple generations and are "prone to perform altruistic acts as part of their division of labor."[11] In the lexicon of evolution altruism means sacrificing one's own reproductive fitness for the benefit of others, and in evolutionary biology reproductive fitness (passing on one's genes) is the bottom line.[12]

In this context the presence of a trait like altruism creates an enigma for evolution because it would appear that those individuals who have this trait would be less likely to reproduce, and therefore altruism should disappear over time. But altruism doesn't disappear and in fact is present in highly social species (insects and humans alike). In insect superorganisms altruism is so profound in the most advanced of social insects (the superorganisms) there are sterile and haploid workers in the colony that have clearly sacrificed their individual reproductive fitness for the benefit of the colony. E. O. Wilson calls this "collateral altruism," which he describes as "behavior benefiting others at the cost of the lifetime production of offspring by the altruist."[13]

This type of extreme altruism by caste does not occur in humans, and for Wilson and Hölldobler it is this distinction that determines the designation of *superorganism*. Indeed, they define the superorganism as an "advanced state of eusociality, in which interindividual conflict for reproductive privilege is diminished and the worker caste is selected to maximize colony efficiency in inter-colony competition."[14] This is specifically what Hölldobler and Wilson mean when they refer to "close cooperation" in insect species. In their words: "The basic elements of the superorganism are not cells and tissues but closely cooperating animals."[15] In other words *superorganism* denotes a unique level of hierarchy in the formation of life. According to this definition humans are eusocial but would not be classified as superorganisms since there is no biological caste of humans that sacrifices reproductive fitness for the group. And there are many species of insects that are classified as superorganisms under their definition that don't cultivate. It is clear that in the case of insect superorganisms there are genetic changes that give rise to their caste system (reproductive and otherwise), which is one of the major components of their division of labor. This doesn't occur in humans.

The presence of altruism moved evolutionary biologists to consider group selection as a player in the matrix of evolution. For many decades the notion that selection takes place at the level of the individual and that each individual has the underlying motivation to pass on genes has been the dominant paradigm and central focus of evolutionary biology. The "selfish gene" has been ascendant in the realm of evolutionary theory. Yet this approach has had difficulty explaining cooperative behavior, especially behavior (like altruism) that appears to work against individual survival and reproduction. The pillars of the selfish gene were expanded through kin selection (you will sacrifice yourself for people related to you) to attempt to explain altruism.[16] But this expansion didn't capture cooperation (and altruism) in large anonymous human societies that were composed of many unrelated individuals.

Group selection has resurfaced to fill this void. It asserts that selection does not simply occur at the individual level but may take place at the group level as well. In other words, there is more than one level of selection at play in the matrix of evolution, and the level of the group may dominate in the tension between the different levels of selection. As you might imagine group selection makes for more complexity in the landscape of evolution. Questions arise of how groups form and when they become a whole with power to dominate the selection of traits. Even E. O. Wilson, once a stalwart of kin selection, has concluded: "The old paradigm of social evolution, grown venerable after four decades, has thus failed . . . kin selection, if it occurs at all in animals, must be a weak form of selection that occurs only in special conditions easily violated."[17]

David Sloan Wilson and E. O. Wilson outline the landscape of group selection as it now stands: "For the social group to function as an adaptive unit its members must do things for each other. Yet, these group-advantageous behaviors seldom maximize relative fitness within the social group. The solution according to Darwin is that natural selection takes place at more than one level of the biological hierarchy."[18] D. S. Wilson tells us that although altruism may give a selective advantage to the individual within the group who isn't an altruist, groups of altruists are likely to out-compete groups with no altruists, and therefore altruism is reproduced. In the words of E. O. Wilson: "An iron rule exists in genetic social evolution. It is that selfish individuals beat altruistic individuals while groups of altruists beat groups of selfish individuals."[19] Altruism, of course, is a simplified case where a single trait is assumed to be at play, so selection works on a "single evolutionary parameter" (in the words of Samir Okasha), but the "trait" is nevertheless reproduced because of the force of the group in selection. Evolutionary biologists have been preoccupied with altruism because it has offered an incisive way to wage the debate between group and individual selection.[20]

Along with group selection, multilevel selection must become part of the lexicon of evolutionary biology because the tension between different levels of selection (group versus individual) must be navigated.[21] E. O. Wilson commented specifically on the importance of multilevel selection with regard to the altruism of insect superorganisms. He tells us that "the existence of collateral altruism is one of the perennial problems of evolutionary biology. Given its genetic consequences, how can programmed sacrifices to collaterally related group members arise by natural selection?"[22] Again, his answer to this question is that the force of natural selection is, in fact, playing out at different levels and the "group" itself has become a force in the process of selection.[23]

There is no question that the ascendancy of group and multilevel selection has been an important step in moving evolutionary biology into a different realm of discourse around the evolution of cooperation. But it is important to further extend that discussion by expanding the understanding of the formation and significance of groups themselves. Analyses about what makes a group a group, and how a distinct group emerges with significance in the landscape of evolution, is still in its infancy. The conversation has thus far revolved almost entirely around a genetic basis for cooperation (e.g., altruism is a trait with an assumed genetic basis). The power of the group feeds back on the selection of genes. And yet the agricultural system can be viewed as an extreme example of the formation of a powerful group and the expression of extreme cooperation that has been very successful in expanding the species that engage it. But the connection between that whole and genetic selection is not clear cut either in its formation or its impact. We certainly don't think of agriculture as being a matter of genetic change in the case of *Homo sapiens*. Humans are not genetically altered in the same way insect agriculturalists are in the formation of the agricultural system. In fact, one can easily argue that humans made the transition to an agricultural system, a new whole, without any significant change in their genome.

Here I raise the possibility that the genetic changes we observe in the formation of insect agriculturalists might be accommodating to (or masking) the force and importance of the ascendant agricultural system. It seems that if we are to seriously explore the emergence of agriculture across species and the similarities we find therein, the aperture of evolution will have to be expanded to consider that the system of agriculture is its own force, and its formation must be considered independent of genetic change.[24] We might miss formidable processes of group formation when we focus on genetic change. And conversely a significant evolutionary change may be missed (in humans, for example) if the barometer of significant change is calibrated on the basis of genetic change. The formation of some agricultural groups (e.g., insects) involve genetic change, while others (e.g., humans) do not. But we know that the expression of cooperation in agriculture is structurally and dynamically the same no matter the species. Agricultural groups become extremely successful in expanding their numbers, and all form a highly integrated productive whole. For want of a better term we can label all the species that engage with agriculture *economic superorganisms*. Again, I am making the distinction here between superorganisms as described and defined by Wilson, Hölldobler, and others and an economic superorganism that refers specifically to agriculture. Some insect species are superorganisms. No humans are superorganisms, but some insect species and most of humanity became economic superorganisms with agriculture—they are bound together in the productive whole of agriculture. Economic superorganisms are joined in a vast enterprise of cultivation affecting both the fitness of the group, the integrity of the individual vis-à-vis the group, and the relationship of the group to the world outside the group. There is no category for economic superorganisms in the lexicon of evolutionary biology as far as I know. For agricultural insects the emergence of agriculture involved genetic selection over a long period, for humans with the power of quick adaptation through culture was their mechanism of engagement. But the power of the emerging agricultural system is as formidable for humans as it is for agricultural insects.

An extensive division of labor around cultivation is one of the most salient characteristics of the form cooperation takes in an agricultural system, and this is true for all species that engage agriculture. Extreme caste formation (a genetic matter) as found in insect agriculturalists may help to facilitate the division of labor, but in a broader sense it is simply one mechanism of extreme role differentiation that accompanies agricultural insects and, in fact, all agricultural species.[25] In humans, the division of labor is enabled by the capacity for culture. Unfortunately, the evolution of the capacity for culture has overshadowed a more expansive exploration of the unique expression of cooperation found in economic superorganisms. It has served to cordon the analysis of humans from other economic superorganisms, and it has diminished our understanding of the importance and power of the agricultural system.

What we can say is that both insect agriculturalists and humans that engaged agriculture had propensities for evolving extreme cooperation through a division of labor that was essential to the formation of a collective material whole; that is, the agricultural system. And the division of labor around the focal point of cultivation reinforced and cemented group formation. The engagement of the division of labor around agriculture was facilitated by culture in humans and by the slow iteration of mutation and selection in insect superorganisms. In other words, multiple mechanisms can be used to facilitate the formation of an extensive division of labor that gives rise to economic superorganisms. But the extension and direction of this capacity through agriculture and the role it plays in the formation of the system as an integrated whole is universal. It is here that the lines between evolutionary process and economic structure and its formation intermingle and perhaps make both evolutionary biologists and economists uncomfortable.

The process of iteration toward the agricultural whole is itself dialectical where the advantages of the division of labor, population growth, and food production reinforce and extend each other in powerful feedback loops and an expansionary spiral with very real implications for fitness and group formation. So while there are dramatic differences in mechanisms of engagement, elaborations of the system, and experience with the agricultural system in different species, in broad outline and final outcome a very similar and powerful system has been formed by otherwise extremely different species. It is also clear that very few species engage in agriculture, but all who do become economic superorganisms. We are left to wonder whether collective configuration (elaborate and materially interdependent division of labor) and its focus (cultivation—food production) and the powerful feedback loops defined therein should shift the focus of evolutionary discourse toward a more expansive inclusion of the formation of integrated and powerful economic systems.

One thing is certain, a powerful economic system was engaged with agriculture that imparted collective fitness to the groups that practiced it even as individual autonomy for those involved was drastically eroded. This is as true for humans as it is for insect superorganisms. It is clearly the case that the boundaries between the individual and the group were altered by this arrangement. Hunters and gatherers might survive independently of the group, but those societies engaged in grain agriculture left little room for individual survival outside of the agricultural system. And if you think about the legacy of this type of system as it is manifest in the present, you can understand that we all feel powerless in a vast system in which each of us plays a bit part. We are existentially interdependent in material life. Perhaps more importantly, the agricultural system is an integrated whole in relation to what is external to it, and what is external to it for humans is the more-than-human world.

It is essential to give at least some background on agricultural insects so that you can more fully appreciate why I use them to draw a corollary to grain agriculture in humans. My goal here is not to elaborate every detail of the specific history of the evolution of insect agriculturalists because there are many entomologists (Wilson and Hölldobler among them) who have done that work.[26] Nor do I want to carry the comparison of humans and insects to an unnatural extreme claiming we are no different than insect superorganisms that practice agriculture. We are different, and I will spend ample time elaborating those differences in the chapters that follow because they are important to the human story—and that story is the focus of this inquiry. But it is important to note an appreciation for other species that demonstrate a remarkable cooperative potential in material life with agriculture.

For those of you not familiar with the amazing agricultural insects let me provide you with a brief description of one of the most highly evolved of the agricultural ants, a group with many species in the genus *Atta* that are commonly known as leafcutter ants.[27] As you might imagine, they cut leaves and process them in assembly line fashion employing both a detailed and social division of labor. They use the leaves to cultivate their fungal gardens. The largest ants cut big pieces of leaves that are transferred to smaller ants who further cut them and so on until they end up with the smallest assembly line ants that "mold the fragments into pellets, add fecal droplets" and insert them in a place where an even smaller ant can plant the "loose strands of fungus" on them.[28] They work with the machine-like precision of an assembly line. An extensive division of labor undergirds the underground colonies they erect, and they are structurally bound together through it. These colonies are architecturally sophisticated and attentive to gas exchange, waste disposal, rearing of young, defense, and fungal production, joining as many as a million ants carrying out their individual roles around their collective enterprise. They have clearly tapped into the benefit of a collective order in their ability to reproduce themselves and expand their numbers around fungal production. They have been actively engaged in agriculture for tens of millions of years. Mueller and Rabeling note that "leafcutter fungiculture indeed represents one of the key innovations in animal evolution." I would say the same for humans around the cultivation of annual grains. It is a key innovation in human evolution.

The insect colonies that practice agriculture display extraordinary phenotypic variation based on task assignment, and some members of the colony also have the ability to move from one task to another based on need. In all examples of the extreme division of labor individual autonomy is essentially nonexistent. No single ant has knowledge of fungal production, but instead knowledge is embedded in the collective, in the way they are functionally differentiated and cohesively connected around the common purpose of fungal production. And there are formidable feedback loops between collective configuration and fungal production, making these colonies profoundly expansionary. Again Mueller and Rabeling tell us: "Some extant leaf-cutter nests are estimated to live for 10–20 years, have 5–10 million workers and maintain 500–1000 football-sized fungal gardens in an underground metropolis occupying the volume of a bus."[29] These colonies seem to have an intelligence, purpose, and collective material order unto themselves. They are not simply insect superorganisms; they are remarkable economic superorganisms.

Clearly agriculture in insects is an example of elaborate cooperation in the material life of the species that practice it, but the same can be said for *Homo sapiens*. They, too, came to express a very elaborate and extensive division of labor around the cultivation of annual grains that went far beyond what they had engaged as hunters and gatherers, and more importantly they became highly interdependent through the division of labor around the focal point of grain production. Though they don't develop reproductive castes, nor do they morphologically differentiate depending on their roles, they do express such an extensive and structural division of labor that they cannot survive independently of the agricultural group.

Our anthropocentric and human supremacist tendency is to think of agriculture as a uniquely human matter resulting from our big brain, our particular aptitude for sociality embodied in culture and its institutional expression, and our particular aptitude for inventiveness in the face of the search for a better life. Scholars in one way or another view the agricultural revolution through this lens. But agricultural insects achieved a very similar economic structure, and the same "success," millions of years ago and in the absence of these unique traits and triggers. The way we think about the economic life of humans and its formation might have to be revised in light of this similarity.

We are told by evolutionary biologists that when the group becomes important enough in the play of evolution we can say a major evolutionary transition has occurred. D. S. Wilson tells us: "When between-group selection sufficiently dominates within-group selection, the group becomes so functionally organized that it becomes a higher-level organism in its own right." D. S. Wilson calls major evolutionary transitions "one of the most important developments in evolutionary biology."[30] It is clear that the world is made up of increasingly higher levels of organization, and imparting significance to a different level of organization simply reflects the reality we observe. Evolutionary biologists recognize the transition from organisms to societies as a major transition, but what about societies to economic super-organism?[31] So far agriculture has been but a sidebar to what are considered major evolutionary transitions. Whether we choose to place the economic superorganism in the realm of a major evolutionary transition is a matter to consider, but there is no question that this alteration in human economic life put humans on a different trajectory in their relationship to each other and to the more-than-human world.

Many evolutionary biologists and social scientists are contributing now to what is called an extended evolutionary synthesis (EES) to account for the shifting boundaries of evolution. The evolutionary biologist Peter Corning tells us: "The most important common property in each of the major transitions is that novel combined effects (synergies) established a new level of complexity and an interdependent "whole" that became a target of differential selection."[32] Corning calls for "a more ecumenical paradigm" to explain the evolution of cooperation and complexity and account for all of the causal agents involved. Clearly cooperative behavior is a complex matter, and the domain of evolution is now to embrace this complexity.

The commonality among different species surrounding agriculture should further nudge the breadth of the study of the evolution of cooperation. An extended evolutionary synthesis (EES) that can more fully embrace the complexity of collective formation and vast "cooperation" and its power in formation of agriculture has the power to shed light on the place of economic systems in the matrix of evolution. As the sociologist Peter Grimes tells us, "Once a complex structure emerges it becomes an active agent in its own recreation."[33] Think of the feedback loops in agriculture where the division of labor enables cultivation (energy production), which further enables a division of labor and thus adds efficiency and enables further cultivation. And in humans the trajectory of this system change has been particularly important and enduring and has been enhanced through cultural and technological embellishment over ten thousand years. The duality created between humans and the more-than-human world is one of the more important results of this system change. That duality has reached a crescendo, as the agricultural system has had ten thousand years to mature.

E. O. Wilson advises that "the evolutionary origin of any complex biological system can be reconstructed correctly only if viewed as the culmination of a history of stages tracked from start to finish."[34] Wilson tells us that "actual histories" of species that exhibit extreme forms of sociality must be mapped out where it becomes clear that the group has become something distinct and the individual is clearly regulated in his/her behavior according to the demands of the group.[35] It is informative to map out actual histories. Yet this should not be done at the expense of attention to cross-species similarities that can broaden the application of the boundaries of evolution in understanding collective formation in species life. The evolutionary importance and commonality of agricultural systems and their power as a whole is overlooked by focusing exclusively on the provincial histories of different species. Insect agriculturalists and humans that began large-scale grain agriculture are clearly engaging something similar.

If we recognize the universality and power of the agricultural system and the processes that formed it perhaps we might move away from human-centric narratives that lead us in the direction of hubris. A more expansive perspective and greater humility in contemplating the economic trajectory we find ourselves with might be advantageous to us now. Are we on a devolution into the economic superorganism or on the road to progress? We know that the human species is moving across a divide where it is overwhelming the Earth with human bodies and their exosomatic extensions. The sixth mass extinction is upon us. Can we navigate our way back, not to a Pleistocene life, but to our Pleistocene roots? Can humans truncate the trajectory of the economic superorganism? Can they once again become an embedded species—one of many? These are lingering questions that still go unanswered.

Chapter Two

Agri-culture?

It is clear that agriculture set the pattern of an economic system that placed humans on a different trajectory around material life. What we know is that an agricultural system is universal, a manifestation of cooperation in material life expressed similarly by very different species. To date there has been little appreciation for this aspect of agriculture, and little effort has been made to see the similarities among the diverse species that practice it. Instead we are left with species-specific analyses and a lack of appreciation for the universality and force of the agricultural system, its unique expression of cooperation and sociality, and any place it might hold in the matrix of evolution.

The discourse around the human transition to agriculture posits no particular evolutionary significance to this momentous change in material life. It is viewed simply as a particular expression of culture where the capacity for culture is thought to be the evolutionarily significant step in the evolution of human cooperation. This parochial approach is mirrored in the case of insect superorganisms. Agriculture is a subset of the evolution of the superorganism. The attainment of eusociality and the status of superorganism are considered the evolutionarily significant expressions of insect cooperation, and insect agriculture is simply a particular form of these.

I have no quibble with the perspective that the human expression of cooperation is unique by virtue of the human capacity for culture, nor that the attainment of culture was evolutionarily significant for humans, a major evolutionary transition. (Nor do I have a quibble with the uniqueness of the biological superorganism in insect cooperation.) Yet it seems that the uniqueness of human cooperation should be explored in addition to, and not in place of, the commonalities that tie our expression of cooperation to that of other species particularly as it takes universal form with agriculture. Agriculture is a universal system tendency expressed in many species, and one of the ways we come to understand what is unique to humans is by exploring what is not. Ask yourself what we've learned about ourselves by studying the behavior of other animals.[1] It is legitimate to ask how one of the most momentous changes in the collective material organization of our species came to be viewed as a remnant of culture rather than an evolutionarily significant change in its own right.

As you may recall from chapter 1, the discussion of the evolution of human cooperation initially focused on altruism (an extreme example of cooperative behavior affecting fitness) but subsequently has come to orbit culture, a more expansive manifestation of the robust capacity for cooperation in humans. Truth be known, culture is a loosely defined term in evolutionary biology, and according to some evolutionary biologists it isn't even clear that culture is solely the domain of humans. For example, Kevin Laland tells us, "Through culture and society, all of us inherit knowledge and skill acquired by our parents. Evolutionary biologists have accepted this for at least a century, but until recently it was considered to be restricted to humans. That's no longer tenable: creatures across the animal kingdom learn socially about diet, feeding techniques, predator avoidance, communication, migration, and mate and breeding-site choices."[2] For this reason, Laland rightly parses human culture more carefully.

According to Laland human beings possess "high-fidelity transmission mechanisms including an unusually accurate capacity for imitation, teaching and language," which are thought to facilitate the cumulative aspects of culture building. Humans are considered to be unique in their capacity for "cumulative culture," which is built over time. Humans also have a unique ability for innovation that builds and expands on previous knowledge and translates that knowledge from one generation to the next in an accurate way. In all these ways human culture is unique.[3] And the advantage of culture is the quick adaptability it imparts on humans.

Scholars who recognize the importance of the agricultural revolution in human history do not consider it to be an expression of cooperation worthy of evolutionary distinction. Rather it is the capacity for culture that is considered evolutionarily distinct. E. O. Wilson refers to agriculture in humans as "a major cultural transition" connecting his analysis to the dominant thread of *culture* in the literature on human sociality and cooperation. Kevin Laland attributes the transition to agriculture to the unique intelligence of humans that he claims is the result of our capacity for culture and social learning.[4] Laland tells us that human intelligence itself developed in a dialectical dance with culture—a process of selection for the rewards of an enhanced and refined capacity for social learning unique to humans. Again among the scholars who explore human cooperation and sociality, the attainment of the capacity for culture is clearly *the* important milestone in the evolutionary history of human sociality and cooperation, agriculture is derivative of this more important evolutionary transition.

Any unique importance of the agricultural revolution is simply lost amidst this clamor around the evolution of the capacity for culture.[5] There is an adage that is best to keep in mind in the context of this discussion: if the only tool in your toolbox is a hammer it isn't surprising that everything begins to resemble a nail. If the uniqueness of human cooperation is viewed through the focal point of culture then every aspect of extensive cooperation (of which grain agriculture is an example) starts to look like an aspect of human culture. It is as if a full accounting of the evolution of cooperation suddenly went silent on the question of the unique and universal expression of cooperation as it is embodied in a universal agricultural system. The consensus among scholars that the attainment of the capacity for culture denotes a major evolutionary transition in humans is not something I disagree with.[6] Nor do I disagree with the proposition that the human attainment of culture, as Laland has described it, is unique. I merely ask whether it is accurate to see the particular manifestation of cooperation expressed in the agricultural revolution exclusively in this light.[7]

Let me return to agricultural insects for a minute. Many insect species exhibit vast cooperation around agriculture that is not a matter of the culture that humans possess, but the structure and dynamic of their cooperation around agriculture looks strikingly similar to the structure and dynamic of human cooperation around agriculture; this again leads to the logic of questioning whether human culture adequately captures the etiology of the agricultural mode of production and its evolutionary significance even in humans. Clearly cooperation and the role of culture in the creation of agriculture needs more careful parsing. Yet human exceptionalism through culture dominates the conversation on human cooperation and by extension the human transition to agriculture.[8] The name itself—*agri*-culture reflects this orientation. It literally means "culture around land." The truth is, of course, that *Homo sapiens* have always had culture around land.

The anthropologists Pete Richerson and Robert Boyd are among the vanguard of scholars who have developed the literature on human culture and its connection to evolution. They have helped to build the framework known as gene-culture coevolution. The idea behind gene-culture coevolution is that humans evolved a social psychology that enabled culture and its transmission through social learning. The capacity for culture (its replication and modification) was advantageous because it acted much like genes (responding to change), only quicker. It is thought that the development of a capacity for culture, as well as its manifestation in technology, was especially important for human adaptation to the unstable climactic conditions during the Pleistocene—in other words that period when we became fully fledged *Homo sapiens* with culture. Richerson and Boyd tell us that humans developed a "tribal instinct" during the Pleistocene through this evolutionary process that allowed them "to interact cooperatively with a larger, symbolically marked set of people, or tribe."[9]

In this framework the social hardware necessary for the development of culture is genetically based, evolving through this dialectical process involving group selection. Hence we became progressively more cultural over time, and culture itself impinges on genetic selection. Herbert Gintis states this eloquently: "We are the species that we are because . . . genes provide individuals with the capacities and incentives to transform culture, and culture guides the transformation of the gene pool from generation to generation."[10] Through this process humans with greater capacity for culture would have greater survival rates. The power of the group is reinforced, and the cultural group becomes progressively more important in the play of evolution. Symbolic markers may extend to "society-specific labels" such as a common language that provide mechanisms to identify members of a group that might not be personally known, thereby extending our capacity for cooperation to those we don't know personally and who are not kin.[11] The formation of institutions that become part of the cultural landscape further encourage cooperation and cement the force of the group.[12] A dialectical dance between genes and culture is at play. Culture becomes the quintessential expression of the human capacity for cooperation. It is essentially a type of niche construction.

Over time the framework of gene-culture coevolution has been extended and refined. Let me return to Kevin Laland who illuminates further the social evolution of humans, extending it to the evolution of human intelligence itself. He argues that the building blocks of cooperation and in particular the ability to copy the behavior of others is found in many species (these basic capabilities don't require "extensive brain circuitry" since a broad spectrum of animals are able to copy and innovate). Copying the behavior of others can be a good strategy for survival because learned behavior can spread, and individuals can save time and energy in employing new strategies by letting someone else do it first. In Laland's words: "Copying pays because other individuals prefilter behavior, thereby making adaptive solutions available to others to copy . . . Natural selection favors more and more efficient and accurate means of copying." But the ability to be strategic and efficient about copying and to employ effective high-fidelity copying is only found in humans who develop the ability for a more "strategic, accurate, and cost-effective strategy to copy." In particular, they can learn from one another and pass on knowledge that is accurate and cumulative over generations. Humans manage this refinement of social learning, and in dialectic interplay it develops their intelligence that further refines their cultural capacities and intelligence.[13]

As previously noted, Laland is clear that individuals in nonhuman species often engage cooperative behavior in "foraging, hunting, and defense . . . but in such societies individuals rarely take up a variety of distinct and coherently integrated roles. That would require some means of coordinating the behavior of the collective, and such mechanisms are generally not present."[14] Again, in his framework the coordination problem is resolved through the particular attributes of human culture and the unique human intelligence that goes with it. Yet the example of insect agriculture indicates that the capacity for taking up a variety of distinct and coherently integrate roles resides in the many species that function as economic superorganisms. Intelligence, culture, and social hardware that humans possess are not required. In fact, Duarte et al. tell us: "Higher cognitive processes are not required to achieve complex group behavior."[15] Insect agriculturalists have the ability to refine themselves around agriculture and to perfect this mode of production as exemplified by the leafcutter ants as previously noted.

The literature surrounding the evolution of human cooperation and its central focus on culture suffers from a humancentric disposition. The human capacity for cooperation becomes solely focused on culture and human intelligence, which eclipses a more expansive look at the complexity of the formation of cooperation, especially in material (economic) life. D. S. Wilson says, "The advent of agriculture enabled us to increase the scale of society by many orders of magnitude through a process of cultural multilevel selection."[16] There is no question that some cultures have an advantage over others, but this extension of multilevel selection to culture further eliminates a more expansive and nuanced understanding of cooperation, especially as it takes form around material life. Any unique significance economic (material) order might have in the matrix of evolution is eliminated since different material systems are simply considered the result of culture.

We might consider that for humans there are evolutionarily significant differences in cooperation (and its role in group formation) that might be understood, not in the light of culture, but in the light of cross-species similarities and the formation of material systems. Culture may be the mechanism used to engage agriculture in humans, but it feeds into something more universal; a structural and dynamic integrity develops around agriculture turning the species that practice it into an interdependent, expansive whole—an economic superorganism. Culture helps to engage the economic superorganism in humans that then reshapes what we commonly think of as culture (ideology, institutions, language) feeding back into the economic superorganism and creating the human agricultural system in its particular totality.

Clearly the challenge is how to go about exploring the particular cooperation that defines agriculture. Members of the human agricultural group are not simply tied together by language and institutional life. They are instead structurally (and existentially) bound together around the focal point of food (grain) production. While food production is always central to the material life of a species, cultivation is a unique approach to this necessity. It is a process of actively producing food in a carefully choreographed process around a focused enterprise. Agricultural systems are profoundly cohesive and expansionary, and the feedback loops that define them are clear and powerful.[17] It is appropriate and indeed necessary to look closely at the universal structural integrity of the agricultural system if we are to appreciate its significance in species life and its significance in human history.

Role partitioning (the division of labor) around cultivation is at the heart of the expression of cooperation in all agricultural societies. Culture may facilitate the engagement of the division of labor in humans, but its system role is not cultural. Without culture eusocial insects had to wait for genetic mutation and selection to facilitate the expansion of their division of labor around cultivation. But the iteration toward an extensive division of labor, regardless of the mechanisms used (culture or mutation), was part of the process that fully engaged the emergent agricultural system. The expanded food and energy produced then expanded the division of labor and population that further expanded energy (food) production, etc. There was clearly a pull in this direction for all species that engaged cultivation. There are universal benefits to the division of labor that interfaced well with the format of the agricultural system.[18] Economists have long understood that there are efficiencies to the division of labor that are universal and undeniable. Biologists studying insect superorganisms note the same. These efficiencies feed back on the selection process for both insect superorganisms that practice agriculture and human societies that engage agriculture and expand the division of labor: thus, this process fuels more efficient cultivation that provides more food and energy. The division of labor around the focal point of grain production is extended but also provides cohesion to the group, thereby engaging a dramatic expansionary and self-reinforcing system. All agricultural societies in humans involved in the production of annual grains develop in the same direction—they are expansionary, extend the division of labor becoming more structurally interdependent and cohesive, and they become increasingly self-referential, standing in contradistinction to the world outside the agricultural system.

Kevin Laland's preoccupation with human uniqueness suggests that only those species that can resolve the coordination problem are able to deploy "a variety of distinct and coherently integrated roles." While this might be true, his assumption is that humans, through culture, are *the* species able to execute this task. He is partially correct. Humans can execute this task. Laland tells us that "through our language, teaching, and the inadvertent construction of learning environments for others, humans can solve the coordination problem; they can assign distinct roles to individuals and ensure each is trained."[19] But the truth is that the coordination problem might be resolved through chemical signals, for example, and the division of labor can emerge without culture through a gradual process of mutation and selection in an interplay around cultivation as was certainly the case in insect superorganisms. Insect superorganisms role partition and control that partitioning in a very elaborate way. Again, take the example of the leafcutter ants who erect vast civilizations with very attenuated divisions of labor without the benefit of language, teaching, or the construction of learning environments. They do communicate, but this is not the same as language. The division of labor is mostly (though not entirely) genetically based for insect superorganisms, so genetic change is the mechanism for engaging the division of labor. And it is true that for at least some of the tasks involved individual insects can switch from one task to another as needed, so they have some flexibility in this way—this flexibility is triggered by epigenetic phenomenon.[20]

The point is that the deployment of a division of labor and role partitioning and coordination need not rely on unique attributes of human culture. One way to look at it is that the division of labor, like social learning, is a potentiality that crosses species boundaries. (Again, more will be said on this in the next chapter.) Agriculture elaborates this potentiality. Just as human culture is a unique expression of the universal tendency toward social learning, agriculture is a unique expression of the universal tendency toward a division of labor. The most extensive division of labor in insect superorganisms are those that cultivate, and the human division of labor expanded many-fold around the agricultural mode of production.

All species become involved in the same powerful system with its powerful feedback loops, and the inclination of this system is toward a particular expression of cooperation regardless of how a species gets there. When the humans-are-unique story dominates the narrative of agriculture, everything is filtered through the lens of culture and human intelligence, and the emphasis on the evolution of cooperation comes to rest entirely around these capacities and with this focus. Humans are unique in their capacity for culture and in their intelligence, but these are only part of the story of the attainment of an agricultural system and the unique expression of cooperation it exemplifies.[21] A more expansive and critical approach to the engagement of agriculture forces us to reassess our understanding of economic life, how it comes to be what it is, and its significance in human history. The economic superorganism is a whole: a cohesive and expansionary system and a powerful one at that. It is a unique expression of human sociality and cooperation, but it is more than that: it is a thing unto itself.

Let's return to evolutionary biology for a moment. Of the agricultural revolution, Kevin Laland claims the following: "More than anything, agriculture required populations to draw on their guile and ingenuity and seize the opportunity by actively creating the circumstances that rendered it economical." He continues: "In early agricultural societies, the pressure to generate sufficient food to feed the mushrooming population demanded division of labor and occupational specializations. For society to function efficiently, relevant skills and expertise would need to be passed among unrelated individuals, but these skills would frequently be far too complex to pick up simply through imitation."[22] In this framework human intelligence and the unique attributes of human culture are responsible for intentionally seizing upon the division of labor to be more efficient in the production of food because of the burgeoning population.[23] Again, we know many agricultural species seized upon the benefits of the division of labor who have neither intelligence or culture. And it worked similarly for all of them—providing efficiency and cohesion that fed back on the ability to cultivate further expanding population and the system that included furthering the division of labor.

As an aside I'd like to set the record straight on the matter of the intelligence required for agriculture—it really isn't clear that agriculture took as much intelligence and innovation as we have been led to believe. Nor is it clear that the jobs associated with agriculture were so difficult to master that they took extensive training and therefore demanded a great propensity for social learning to extend and orchestrate. The surplus from agriculture did create occupations that might have required more skill, and more skill might have been required to manage surplus, which made society more complicated; however, grain agriculture itself was certainly not highly demanding in inventiveness nor the skills required. The day-to-day of agriculture was more accurately simply repetitive and quite mundane. And the miracle of human inventiveness amounted to being able to observe a seed germinate and seek out the opportunity of wild stands of grains and extend that opportunity. How hard could this be for a species that had evolved to be observant if not scientific?[24]

There is a reason that agriculture was a good fit for capturing and extending the division of labor, and it is this: much of the work associated with agriculture was easily routinized and rationalized, and this made it a good candidate for deploying and extending a division of labor that was more integrated: if there's one thing a division of labor around a task thrives on, it's some regularity. In fact, agriculture clearly worked somewhat contrary to the intelligent interchange with the more-than-human world that humans had mastered up to that point where executive decision making and assessment on an individual basis was a standard part of the provisioning of day-to-day material life. The deskilling associated with agriculture is a fact pointed out very clearly by James C. Scott, who tells us: "I am tempted to see the late Neolithic revolution, for all its contributions to large-scale societies, as something of a deskilling."[25] And it is an idea reinforced by Yuval Harari who has labeled agriculture "history's biggest fraud" precisely because it wasn't a matter of human intelligence. In his words the "responsibility for agriculture lay in the hands of plants. These plants domesticated *Homo Sapiens*, rather than vice versa."[26] The knowledge and executive decision making required to live by hunting and gathering were legions more sophisticated (and more interesting) than that required by either the invention or execution of grain agriculture.

It is quite possible that what we consider to be the tipping point of human cultural sophistication that supposedly led to the attainment of civilization and progress, was actually a tipping point of a different sort. It is also possible that with this change humans were set on a path away from the evolutionary trajectory of a bigger-brained, observant species that they had become up to that point. From the time of the agricultural revolution forward humans would become on average less intelligent and observant, and one could argue their capacity for culture would be more extensively hijacked for purposes of accommodating the structure and dynamic of the economic superorganism. In the standard narrative of human cooperation enabled through culture, groups of preagricultural hunters/gatherers differ from large-scale agricultural societies only by cultural variation, so we need look no further than culture to understand agriculture and its place in our history. It is simply the scaling up of a "tribal instinct" attributable to the gene-culture coevolution that occurred previous to agriculture as Peter Richerson and Robert Boyd claim, or an expansion of the capacity for social labeling that in humans allowed for "the transition to larger agricultural societies" as Mark Moffett points out, or a unique coordination of diverse tasks only enabled by culture as Kevin Leland would have it, or an extension of gene-culture coevolution, which, as Herbert Gintis tells it, is simply a special case of "niche construction."[27] In short, this narrative paints agriculture in the image of the uniqueness of humankind, rather than to consider that agriculture was a point where humankind took a turn toward the economic superorganism not unlike the insect superorganism, which employed essentially the same mode of production.

There is no question that preagricultural humans attained a level of sociality not obtained by other primates nor by other species. The question entertained here is whether that sociality (capacity for cooperation) found its way into agriculture as a categorically different expression of cooperation. Should the agricultural group be considered a different whole in the matrix of evolution, like the movement from organisms to societies? Was agriculture brought about by culture and the unique capacities of human cooperation? Or was culture hijacked in dialectical interplay to facilitate the engagement of an economic superorganism that actually worked against the continued evolution of human intelligence and violated the best impulses of cooperation to live in ecological community? These are questions worth pondering.

It is not unusual in evolution for a particular trait or group of traits to be co-opted for a different purpose. In humans, dramatically different expressions of cooperation may not involve changes in the genome, but that doesn't mean there isn't evolutionary significance to these differences nor that they shouldn't be viewed through the light of evolution.[28] An altered expression of our genetic and epigenetic potential in our collective formation in material life (what I like to call our collective phenotypic expression) may be as important as any change in the genetic composition of our species in the unfolding evolution of humans and of life on Earth.[29] And in humans, the breadth of phenotypic expression both individually and collectively is wide and can become structural and enduring.

The alteration brought about by the agricultural revolution changed the relationship of humans to each other, to the more-than-human world, and it altered the energetics of human material life. Surplus and expansion in material life, duality in the human relationship to the more-than-human world, and profound, almost mechanistic interdependency in material life became the order of the day beginning with grain agriculture. *Homo sapiens* became disarticulated from the rhythm and dynamic of wider ecologies and instead engaged a self-referential dynamic of ever-expanding human population, interdependence, and grain production. Yet, I daresay were we to look into the genetic composition of individual humans, the alteration of humans through grain agriculture might produce barely a whisper of change. It is, however, producing a profound change in the aggregate genetic endowment of life on Earth. That genetic endowment is becoming mostly of the *Homo sapiens* stock eliminating much of the "collective knowledge" of Earth's history. And it does indicate that humans have engaged a strategy for survival that has been (at least temporarily) extremely successful in the measurement of their own fitness (while possibly assuring long-term collapse).

Cooperation can be expressed in a broad range of interpersonal activities including putting arms around each other and chanting incantations, or collective hunting; or it can be expressed in impersonal activities like engaging in an extensive division of labor to facilitate large-scale agricultural production or to build pyramids or cars on an assembly line. It can even include phenotypic differentiation in superorganismic ant or termite colonies or can be expressed in a common "capacity for symbolic thought" such as language. *All* of the above are expressions of an evolved cooperation. Yet they are clearly functionally and structurally different, and they arise uniquely in the unfolding evolution of species. We need more refinement if not evolutionary deconstruction of cooperation in light of the agricultural revolution.

Whether humans can step back from agriculture's legacy of expansion, interdependence, and the duality created between them and the more-than-human world remains to be seen. Can humans undo the whole that has been created and amplified over the past ten thousand years? Is culture the adaptive mechanism it once was for humans? Or in the face of exponential flight has it become a lag on fast change? The altered strategy of material life that began with agriculture has had such force that it has guided the formation of institutions, ideologies, and innovation practiced by humans in the ensuing ten thousand years and not the other way around. I realize my argument has the unfortunate tenor of determinism for the social scientist and perhaps not enough determinism for the evolutionary biologist, but any account of economic order is always caught in this space. Material systems (economic systems) of species are the product of dialectics in evolution and the complexity of emergence systems played out on planet Earth. Of course, culture is involved for humans but so are systems organization, coevolutionary and dialectical processes, and the boundaries and imperatives imposed by structural relationships in material life that include its energetics, interdependencies, and efficiencies. What we can say without equivocation is that the agricultural system moved *Homo sapiens* away from being an observant mammal using "nature as measure" in the execution of our daily life. It institutionalized expansion and hierarchy, forged a structural duality between humans and the more-than-human world in material life, and created a profound interdependence in material life that reduced individual autonomy making humans a different material whole. If one could see ahead even a hundred thousand years, we might look back at our history and conclude that we crossed a great evolutionary divide with the transition to grain agriculture where we became *Homo sapiens agriculturii*.[30]

Chapter Three

The Division of Labor

The proposition put forth here is unsettling. Human societies began to function much like an economic superorganism when they engaged agriculture. Here the level of integration and interdependence in material life became sufficiently profound that individual autonomy was eclipsed by the demands of the economic system. Grain agriculture reshuffled the deck of human sociality and the expression of human cooperation and a dramatic alteration in the structure and energetics of human economic life arose. This was a very effective fitness strategy that essentially moved a large mammal from a k-selected species to a high reproduction and high death rate species in a very short time. Children with a biological clock for slow maturation were cut short in their climb toward adulthood under the regimen imposed by agricultural work and later by its derivative system—capitalism. With the capacities of humans for both institutional and technological elaborations (cumulative culture) humans extended the powerful economic system and entered an ecologically destructive phase of their collective relationship with Earth. They also entered a different relationship to each other as they moved toward expansion, surplus, and hierarchy.

The genetic endowment humans arrived with at the door of the Holocene (an endowment honed over eons of time and refined in the Pleistocene) was expressed in a different way as the purpose and context of human economic life was altered with grain agriculture. The oddity of the agricultural system for humans is that in many ways the agricultural mode of production was not particularly complementary to many of the inclinations of the humans that practiced it (although we make no such assessment with regard to agricultural insects). For humans it was backbreaking, provided a rather poor diet, increased disease, made for a lot of horribly routinized and standardized work, reducing executive decision making in day-to-day living. One can hardly claim it was a good use of human intelligence on average.[1] It was only a good use of human intelligence for those who were able to live on the surplus provided by others who toiled. Most importantly, the engagement of grain agriculture created a structured duality between the economic system and the more-than-human world. It was a duality that had not previously existed.

One can say with some confidence that humans didn't get up one morning and say "gee—let's invent agriculture" to make life easier. Rather, the walk through the door to agriculture for humans was mostly unintentional, the outcome of the complex evolutionary dialectics at play in the presence of human potentialities and those of the grains they cultivated. Human culture and inventiveness were accommodating to the emerging agricultural system, and they have mostly remained accommodating in the ensuing ten thousand years. To think about the agricultural revolution and the engagement of the economic superorganism in this way moves the understanding of the economic life of *Homo sapiens* into a deeply materialist realm and the etiology of human economic order moves away from a purely humancentric perspective.

Our cross-species exploration of agriculture points to one particularly important species proclivity that stands out in the emergence of agriculture and that is the capacity to extend the division of labor. This is a unique aspect of cooperation that helped to form the structure and cohesion around cultivation. All species that engaged agriculture extended their respective divisions of labor once they began the practice of cultivation although their mechanisms of engagement were different. For insect cultivators it was a slow iteration of mutation and selection around colony based algorithmic rules that unfurled the division of labor. For human cultivators it was the flexibility embodied in the human capacity for culture that was a quick mechanism of engagement. But regardless of the mechanism of engagement once the division of labor began to form around cultivation the agricultural group became a formidable cohesive whole and agricultural species entered an evolutionarily distinct realm of cooperation as an economic superorganism.

Thus, the division of labor stands front and center as part of the essence and integrity of an agricultural system and its emergence. Agriculture is a system of food *production*, and like all production it requires a certain organization to execute. Many economists have long identified the division of labor as a central tenant of economic life, although for most it has been viewed as the exclusive domain of humans. In an agricultural system cultivation, the division of labor, and population growth are autocatalytic variables essentially reacting with and extending each other in a positive feedback dynamic. They build on each other in a nonlinear way forging a unique system, a collective whole, and a powerful collective strategy for survival. The group, through a division of labor, engages agriculture. It is accurate to say that the species engaged in an agricultural system become involved in a unique energetic dynamic; however, reducing the system to energy does not capture the collective formation, integrity, and dynamic of the agricultural system.

All species that engage agriculture are changed in similar ways through the process of becoming agricultural whether they be ants, termites, or humans. And they all come to the possibility of agriculture with the potential to be collectively altered according to the demands and dictates of the pull of the agricultural system. Thus, there is a dramatic dialectical coevolution between cultivars and cultivators that is universal. Coevolution is not a one-way street. For the cultivators the formation of the agricultural group through an extensive division of labor is key. This may seem inconsequential to the undiscerning eye, but it is of the utmost significance in group formation, cohesion, and the ensuing dynamic. The cohesion provided through the division of labor as well as the dialectical interaction of the division of labor, energy production, and population growth are essential in understanding and appreciating the significance of the emergence of the agricultural system, a wholly distinct economic system. Clearly, strikingly different genotypes can give rise to what is essentially the same collective economic phenotype as human and insect agriculture reveal. This despite obvious differences in the intelligence of the species at an individual level and very different mechanisms of engaging collective life around cultivation (culture, or mutation and selection).

In the case of humans, the coevolution was primarily with annual grains and in the case of agricultural insects with the fungi they cultivate. I'm not sure of all the ways fungi were changed in this process, but we know that in the most sophisticated fungal production by the leafcutter ants the fungi no longer have the potential to cross over to independent life. Perhaps more importantly, insect agriculturalists developed a much more elaborate and extensive division of labor around fungal production, so clearly they were collectively altered in this way through agriculture. (Please refer back to the description of the elaborate division of labor in leaf-cutter ants introduced in chapter 1.) Nonagricultural species of insects and humans do have a division of labor, but this propensity, which clearly had advantages in many contexts, was universally extended in agricultural species becoming an essential part of the structure of agricultural production.

A similar coevolution occurred with humans. Annual grains provided good coevolutionary material for humans because they were quick to give coevolutionary results by virtue of being planted every year. Any attribute in them that worked well (e.g., nonshattering seeds and large seed size) would be enhanced in a relatively short period.[2] And annuals were readily abundant in strategic locations when the Holocene warming began.[3] The importance of annual grains can't be overemphasized in our understanding of the agricultural revolution in humans. (I will discuss this more extensively in the next chapter.) Agriculture begins after the Holocene warming *independently* in the Levant, Asia, and the Americas; in all cases annual grains are involved (wheat, rice, corn, and eventually millet and barley). This would certainly suggest a species tendency in the direction of cultivation. Yet humans did not simply cultivate crops without being changed in the process just as insect agriculturalists were changed in the process of cultivating fungi. Again, the way to understand the significance of the change is to understand that it was first and foremost a change in the *collective configuration* of humans in material life, and one essential aspect of this collective change is that an elaborate and interdependent division of labor develops around the focal point of food production. The human genome changed very little with grain agriculture, but the collective configuration of humans around annual grain production was distinctively different.[4] It is in this latter sense that there is no difference between insect agriculturalists who cultivate fungi and humans that cultivate grains.

Genetic differentiation may define the superorganism in insects, but it is a mechanism of engagement in the emergence of the insect economic superorganism.[5] In the same way, we might view culture in humans as an essential attribute of their ultrasociality, but it is a particular mechanism for the formation of their economic superorganism. The agricultural system iterates all species in this same direction regardless of their idiosyncratic mechanisms of engagement. Niche construction is not simply a matter of a species changing the external environment in which they reside but also a matter of changing themselves collectively.

Not surprisingly, the division of labor and, by extension, self-organization have become important topics in the study of the evolution of cooperative insects because those who study insect colonies clearly recognize that it is a foundational characteristic of all superorganisms and greatly elaborated in agricultural insects.[6] Nowak, Tarnita, and Wilson view the division of labor as an emergent trait in eusociality, and Hölldobler and Wilson allocate major chapters in their book *The Superorganism* to a discussion of the division of labor, elaborating on the most advanced of the insect superorganisms, the agricultural insects.[7] In the words of Hölldobler and Wilson: "The trend toward still larger, more complex societies has been accompanied by a hardening of the mechanisms that differentiate worker subcastes and labor roles. The most extreme such diversification is achieved in the minority of species that form physical castes. Within the physical castes are folded more finely differentiated physiological castes."[8] The leafcutters became so sophisticated that they are polymorphic.[9] Accompanying this change is also larger colony size. As Mark Moffett tells us, "Leafcutters grow and harvest their fungi using farming techniques no less complex than ours . . . The invention of agriculture has enabled societies of humans and leafcutters, which were farming long before people, to support massive populations."[10] Indeed the population and nest size of leafcutters is legendary, and the alteration in human population dynamics after agriculture profound.

Some scholars recognize the division of labor "as a significant commonality in the organization of insect and human societies."[11] Tim Flannery makes this connection in his review of Hölldobler and Wilson's book *The Superorganism* when he says: "Clearly, not only did the attines beat us to agriculture, but they exemplified the concept of the division of labour long before Adam Smith stated it."[12] Even so, few have seriously explored it as a commonality integral to the structure and dynamic of an agricultural system and a central feature of the economic superorganism. John Gowdy and I began this exploration in our work on human ultrasociality.[13] What is very clear is that humans came to the Holocene with a propensity for a division of labor already established just as there exists a propensity for a division of labor in nonagricultural insects that attain caste status and become superorganisms. In humans the division of labor was engaged rather modestly and loosely before agriculture, mostly according to age and gender. Even so, individual humans were still self-reliant in food provisioning. That is not to say they didn't share food or benefit by doing so, nor that some didn't hunt together, but simply that each adult person had extensive ecological knowledge and the skill base that enabled individual survival. Each person could fend for themselves and day-to-day production of material existence was not mostly a collective enterprise. Also individual humans had a well-evolved capacity for culture by the onset of the Holocene warming. Thus they had the plasticity to become many things both individually and collectively.[14]

We can appreciate the universal benefits of the division of labor in addition to its integral role in cementing interdependence around cultivation. One of the most salient of the universal benefits of the division of labor is the efficiency it provides. The deployment of a division of labor allows for benefits from specialization, but it also allows for saving time moving from one task to another and for different tasks to be performed simultaneously when needed. The division of labor is also beneficial when tasks can be standardized or organized in assembly line fashion. In all cases the division of labor adds efficiency that is of collective benefit.[15] And the division of labor creates possibilities for a species that would otherwise be absent. It is hard to imagine engaging greater social complexity without an extensive division of labor.

Indeed, the efficiency benefits of the division of labor have been recognized by economists for centuries and more recently by entomologists. Almost 250 years ago Adam Smith made this observation: "The greatest improvement in the productive powers of labour and the greater part of the skill, dexterity, and judgement with which it is anywhere directed, or applied, seem to have been the effects of the division of labour."[16] Plato, some thirteen centuries before Smith, had similarly recognized the productive benefits of having individuals specializing according to what they were most fit for.[17] Entomologists who study the evolution of social insects recognize the efficiencies in the division of labor as one of the keys to their self-organization and part of the selection process that imparts the colony with fitness.[18] The centrality of the division of labor in the realm of economic analysis and evolutionary biology is not coincidental. In both cases it is recognized as a core aspect of collective life. The question for evolutionary biologists is whether the collective that forms constitutes an evolutionarily distinct whole. But an equally important question emerges in the realm of economics—does this whole impose a duality between economic order and Earth?

The inclination toward efficiency found in the division of labor interfaced well with agriculture partly because fungal and grain production translated into predictable patterns of production. Agricultural enterprises are inherently inclined toward jobs being done simultaneously as well as sequentially in lockstep. And there were ample opportunities to benefit from specialization in the deployment of agriculture. Think about defense and its role in agricultural societies. And the expansive nature of agriculture and the surplus it creates certainly leaves more possibilities to extend the division of labor as population grows around the surfeit of agricultural output. It is difficult to imagine large-scale insect superorganisms engaged in fungal production without an extensive division of labor, as there are many different tasks that must be done in tandem, sequentially, and in lockstep. Mark Moffett comments on the production process in the leafcutters: "A leafcutting factory might have been the envy of Henry Ford: different workers collect, transport, and mince foliage, apply it to a garden, and eject its decay remnants in an orchestrated flow of material from environment to nest and back out again."[19]

It is similarly impossible to imagine the emergence of large-scale agricultural societies in humans without the extensive deployment of a division of labor, certainly this is true once large-scale agricultural societies developed as they did very quickly after the cultivation of grains took hold. In the case of humans think in simplistic terms—planting, cultivating, harvesting, storing, irrigation, and the many opportunities for a more detailed division of labor within each of these activities. In addition, there was defense, expanded reproduction, and the specializations emerging out of increasing complexity and hierarchy. (I will elaborate the human case further in the next chapter.) Once the process starts the division of labor becomes part of the collective structure of the species and an essential element in the system; all are dependent on grain and fungal production, and the system builds on itself with self-reinforcing feedback loops.[20]

It is one thing to recognize the efficiencies in the division of labor, it is another to see it as an integral part of the way an economic system is held together. Economists clearly recognize the importance in terms of efficiency, but they also recognize the importance of the division of labor in this latter sense—as an integral part of a system. Let me take you back to Adam Smith and Karl Marx to illustrate my point. Adam Smith begins his tome on capitalism with three chapters on the division of labor. Smith not only highlighted the efficiency and productive benefits of the division of labor: he also saw the division of labor interacting with self-interest and markets in a way that made for an integrated system. He opined that there were productive benefits in specialization, and everyone (out of self-interest) could take advantage of them through trade. There was then an intimate connection between self-interest, the propensity to "truck, barter, and exchange" and obtaining the benefits of the division of labor. The material well-being of a society would thus be elevated through the expansion of markets, which also meant an expansion of the division of labor. The point is there was a system that was being engaged, and the division of labor was essential to that system. Again, there is much Smith got wrong in his analysis, but understanding that the division of labor was part of a dialectic system process around markets was not one of them.[21]

Karl Marx recognized that very different economic systems evolved over time and could be distinguished by their social relations. What he said was this: "Each new productive force, insofar as it is not merely a quantitative extension of productive forces already known (for instance the bringing into cultivation of fresh land) causes a further development of the division of labour." He elaborated this point further: "The production of life, both of one's own in labour and of fresh life in procreation, now appears as a double relationship: on the one hand as a natural, on the other as a social relationship. By social we understand the co-operation of several individuals, no matter under what conditions, in what manner and to what end. It follows from this that a certain mode of production . . . is always combined with a certain mode of co-operations, or social stage, and this mode of co-operation is itself a productive force."[22] He clearly recognized an interplay between an emerging system and the social relations. His focus, of course, revolved around the class distinctions of capitalism and the extraction of surplus value at the point of production, but he understood that the division of labor made humans more productive and created "interdependence of the individuals among whom the labour is divided."[23]

There is a history of economists exploring the division of labor and its role in an economic system, but there has been no attempt among them to engage this tendency more universally, nor to connect the role of the division of labor in the emergence and expansion of the universal system of agriculture.[24] Adam Smith believed the human capacity for a division of labor "is common to all men, and to be found in no other race of animals."[25] This was not simply because of his limited knowledge of insect superorganisms and cultivators but because he was focused on the connection between the division of labor and the propensity to truck, barter, and exchange. It was the interplay of markets and the division of labor that had captured his attention.

The sociologist Emile Durkheim was perhaps most insightful about the division of labor when he said this over a century ago: "It is no longer considered only a social institution that has its source in the intelligence and will of men, but a phenomenon of general biology whose conditions must be sought in the essential properties of organized matter."[26] It is clearly integral to an economic system creating efficiency, interdependence, and performing a role in the dialectics of economic formation. The lines between the economic and the evolutionary become less clear when we begin to explore the division of labor. When material activity is focused, standardized, and routinized and when the activity provides an interplay between food production, population growth, and the division of labor as in the cultivation of fungi (in insects) and annual grains in humans there appears to be an especially dramatic and reliable outcome and the division of labor as a species propensity becomes part of the structural integrity of the system and a core participant in its feedback loops.

The evolutionary biologist Peter Corning refers to the division of labor as the paradox of dependency. He tells us: "There is a deep paradox involved in dividing up and sharing the elements of a job. It creates an interdependency; everyone must do their part or the desired outcome will not be achieved." The division of labor creates "a built-in enforcer for cooperation." John Gowdy and I claimed the same in our work on ultra-sociality.[27] When the axis around which the division of labor rotates is the focused production of food, which is also to say the focused production of energy, a very powerful feedback mechanism is engaged. Participation is as essential to individual fitness as breathing. Individuals can't produce alone, and if they try to disengage from the system they will not survive. Such a system becomes increasingly self-referential, essentially reinforcing itself in its feedback loops and cohesion.

Complex systems arise in the reproduction of species that have a power and dynamic of their own, and agriculture exemplifies this fact.[28] What is absolutely clear is that once a move was made in the direction of agriculture humans did not turn back (nor did the insects that practice it). From that time forward the broad outline of human economic life followed a predictable path. Humans engaged an expansive, interdependent, and surplus-seeking system, undermining the sense of community with Earth and establishing a duality between humans and the more-than-human world. The ensuing ten thousand years have been an institutional and technological refinement of this trajectory, which has brought us to a divide in our evolutionary history. On the one side is our Pleistocene past and on the other, the trajectory of the economic superorganism.

Part II

Bitter Harvest

Chapter Four

The Tapestry of the
Universal and the Particular

Albert Einstein once said, "We should make things as simple as possible but not simpler." This is the challenge of the human story of agriculture. There are many layers of complexity, and within each layer there are many paths of inquiry. The cross-species comparison helps to cultivate an appreciation for agriculture as a universal system; the result of universal processes and complex synergies of collective evolution. There is an undeniable integrity, cohesion, and energetic motion to an agricultural system that is common to all species that practice it. Thus the human story is most accurately written in the shadow of these universal processes and tendencies and with an appreciation for the force and integrity of agriculture as a system.[1]

Yet it is essential to navigate the distance from the universal to the particular human story because humans are so clearly different than the many other species that cultivate. Humans did not engage the agricultural system simply because of their unique intelligence, inventiveness, guile, and their particular capacity for cultural life (with its technological and institutional manifestations), but these attributes were not irrelevant either. Humans gave their unique imprint to the economic superorganism through their quick mechanisms of engagement and their tools of elaboration. To add to this complexity, the engagement of the system of agriculture in humans was also informed by chance circumstances particular to human history but having nothing to do with humans, most notably: the Holocene warming, the presence of annual grains, and the abundance of soil carbon found in post-Pleistocene soils. These external conditions must be added to the tapestry of the complex human story.

It is therefore the confection of the universal and the particular that forms the human agricultural system. Humans became a self-referential and profoundly interdependent species of expansion and surplus with agriculture just as their insect counterparts had, but humans did so with their own imprint and their own history. In the end the complexity of culture and ingenuity, coevolution, chance circumstances, and universal processes and inclinations formed a powerful human agricultural system that reframed the trajectory of human economic life and recalibrated the relationship of humans to Earth. This legacy is the backdrop of the present war between economy and Earth.

Much has been written about the rise of civilization in the wake of the agricultural revolution. This is a well-worn story that I don't intend to emphasize. Instead the approach provided here offers a counterbalance to the narrative of progress and human supremacy. I recognize that agriculture brought humans civilization—mostly a benefit to the few that flourished in the many advantages of being on the receiving end of its surplus. But my homage is to the majority of humans who were enslaved directly and indirectly in relentless sweat, toil, and alienation in their daily lives through the agricultural system and its legacy; to the Earth disrupted, interrupted, and temporarily diminished in its established cycles and ecologies; and to the foundational and sacred connection between humans and the more-than-human world that was undermined by the economic superorganism and its elaboration. This counternarrative provides historical context to our present war between economy and Earth allowing us to more fully appreciate the enduring power of an economic system and the way it frames the relationship between society and Earth.

There is some consensus that agriculture was not an unequivocal gain for humans.[2] For example, there is a general consensus that as a result of agriculture human health deteriorated, lifespans were reduced, enslavement increased, the prevalence of disease rose, ecological devastation in the form of soil erosion and loss of fertility ensued, and the probability of ecological and societal collapse intensified.[3] Jared Diamond claims agriculture was "the worst mistake in the history of the human race."[4] Yuval Harari tells us in no uncertain terms that the tale of human intelligence unlocking nature's secrets in the form of agriculture and abandoning "the grueling, dangerous and often Spartan life of hunter-gatherers . . . to enjoy the pleasant, satiated life of farmers" is pure "fantasy."[5] And the literature on the ecological problems of grain agriculture from its inception abound. In his historical overview Paul Sears writes: "Wherever we turn, to Asia, Europe, or Africa, we shall find the same story repeated with almost mechanical regularity.

The net productiveness of the land has been decreased. Fertility has been consumed and soil destroyed at a rate far in excess of the capacity of either man or nature to replace."[6]

Yet few of these critiques focus on the altered economic system that was created with agriculture: a system of surplus, expansion, and profound interdependencies around the focal point of grain production; a system functioning in reference to itself and establishing a structural duality between economy and Earth. Wes Jackson calls agriculture "the fall" and tells us we now have both a problem with agriculture (our present industrial agriculture) and the problem of agriculture (the altered relationship of humans to Earth that began with the agricultural revolution). He is correct. The problem *of* agriculture must be understood with a focus on economic system formation. Grain agriculture was not simply a different way of procuring food; it was a major transition in the order of economic life.

In the face of the problematic outcomes of agriculture it is fair to ask: why would a smart species make this transition if the downside was so apparent? That question can be answered by understanding that system formation is not dictated by conscious decision making. The downside of agriculture was not fully understood as the agricultural system got going, nor was the fact that a profoundly altered relationship between society and Earth was taking form. The "decision" to engage agriculture was not a cost-benefit analysis where humans calculated what they were getting into a priori. Initially humans simply augmented hunting and gathering with a bit of grain harvesting in the context of the Holocene warming when stands of wild grains were abundant.[7] And the benefit from harvesting productive stands of wild grains and extending them to cultivation was immediate, but the downside lagged. The engagement of humans in agriculture happened in small incremental steps with enough boosts and beneficial feedbacks, and enough of a lag in the downside of loss of soil fertility, erosion, human health, and ecological problems that the system took hold before any assessment could be made about whether it was a good idea. Once it started, humans were carried along by the emergence system with its powerful feedback loops that were enhanced by particular human elaboration. Yuval Harari claims that once engaged we were stuck with this mode of production because it was impossible to get off the agricultural treadmill—there were simply too many of us. He, of course, is speaking of one of the many feedback loops (population growth) that reinforced the nascent system.[8]

This interpretation of the engagement of agriculture is in contradistinction to our notion that human intelligence came to the rescue of distressed humans confronting scarcity and moved humanity in the direction of progress through agriculture. No profound human intelligence was required as previously noted, yet this is a common misperception about agriculture—that it had to have been driven by human intelligence (as well as the human capacity for culture). For example, according to the economist Robert Heilbronner, "It staggers the imagination to think of the endless efforts that must have been expended in the . . . discovery of planting seeds."[9] To an observant species it simply wasn't that difficult. Steven Pinker portrays agriculture in much the same way when he tells us agriculture is the "elixir with which we stave off entropy" through knowledge thereby changing our destiny.[10] There is no question that a powerful system dynamic was established with agriculture that changed human destiny, but the story of agriculture is more complicated and less intentional than we have come to believe. To understand it we are forced to think more expansively about the complexity of social evolution, as well as the place of unique human propensities and capacities in the creation of this powerful material system.

When agriculture began it wasn't apparent that human economic life and evolutionary history were being fundamentally transformed. Yet a powerful and novel system dynamic was taking hold where humans went from *Homo sapiens sapiens* to *Homo sapiens agriculturii*, a member of the economic superorganism, in a relatively short period. The system began slowly, but over millennia it picked up steam.[11] In the telling of the story of agriculture it is paramount to avoid the mistake of ranking causality or reducing agriculture to human intelligence and intentionality. Instead the threads of causality must be woven together into whole cloth (as dialectic interplay), and then a new system must be recognized and appreciated as something significant and distinct.

What of the threads of this tapestry? The early boost to the agricultural system might be explained by external factors that nonetheless helped to engage the system. These were particular to the human experience but cannot be attributed to humans. The Holocene warming is the most obvious example. There is general consensus that it provided the warming and stability of the climate that opened the door for grain agriculture. Without the Holocene warming it is unlikely agriculture would have taken hold.[12] As well the largesse of carbon that lay untapped in the Pleistocene soils after the Holocene warming began was also important. Again, this was external to humans but nonetheless helped to pave the human experience. Wes Jackson refers to soil carbon as the first of the five pools of carbon that humans would eventually exploit.[13] We also know that agriculture took hold initially in river deltas where the annual flooding replenished the nutrients needed. This natural fertility cycle of river deltas was something humans could take advantage of, but they didn't create it. Eventually, of course, with more dramatic human intervention around irrigation, agriculture was able to expand its domain into more arid regions.[14]

Aside from these external factors there was a dramatic coevolutionary dynamic between humans and annual grains that can't be overlooked and, in fact, must be central to our understanding of the emergence of the human agricultural system. Inherent qualities in both human beings and the grains they cultivated played off each other. Species potentialities are the result of the long arc of the evolutionary history of each species, and it is essential to understand the nature of both humans and annual grains and the interaction between the two. Clearly humans don't have the capacity to photosynthesize, but they do have the capacity to increase the photosynthetic production directed toward themselves through cultivation. This supplies them with both nutrients and energy. While humans accessed the stored carbon in the soil with agriculture they also became active agents in the collection and conversion of solar energy into a usable form for human metabolism. Unique human propensities enabled this process—for example that autochthonous potential for division of labor (reinforced through culture), intelligence, as well as the institutional stanchions that enhanced the evolving agricultural system. Although grains were certainly not the only plants that humans cultivated, they were the determinate cultivar in the engagement of the agricultural system (and the making of civilization).

Annual grains also had inherent attributes that interacted with human potentiality, pushing agriculture to become the system it became.[15] We know annual grains provided good coevolutionary potential because they were quick to give good coevolutionary results by virtue of being planted each year. (In this they were a good analogue to the inherent potentialities of humans who could select and plant based on characteristics that fit human need. Not all selection was intentional as previously noted.) The fact that annuals are annuals meant that the offtake from their cultivation had the potential to vary dramatically from year to year. This variability in production, coupled with the fact that grains can be stored, pushed the envelope of maximum production for any single year as a way to guard against the vagaries of annual production. In this most basic way the envelope of surplus and expansion was pushed reliably outward with annual grains.

The expansionary push of annual grain cultivation also depended on the particular ecology of grain production. While annual grains had tremendous coevolutionary potential and gave quick selective results, they were ecologically challenged. Ironically, the ecological problems inherent to annual grain cultivation also pushed the agricultural system in the direction of expansion and surplus. Part of the ecological challenge stemmed from the fact that annual grain cultivation was (and is) a terrible option for soil. Annuals are evolved to inhabit disturbed areas, and annual grain production required (and requires) continual disturbance of soil that is both depleted and eroded through their cultivation. Efficient planting of annual grains typically began with the removal of soil cover because removal helped to limit competition and added to the efficiency of grain production. Wes Jackson tells us that "agriculture was the first simplifier on the landscape, and with it species extinction and extirpation were huge."[16] Because annuals are harvested and replanted every year the soil is left exposed, further adding to the problem of erosion.

Expansion of agricultural production onto new ground was one of the countervailing strategies against soil erosion and loss of soil fertility (integration of animal and human waste into this balance helped to recycle nutrients and mitigate against the loss of fertility). The expansion of population and urban life (one of the feedback loops of agriculture) created further ecological problems with soil erosion, as deforestation to build urban centers led to siltation of irrigation systems and problems with watersheds.[17] And soil erosion and other ecological problems became progressively worse with the technological proficiency of humans to enhance annual grain production through plowshare and irrigation, for example. We know for example, irrigated fields (a technological innovation to expand agriculture) created the problem of soil salination. This shifted the agricultural mix toward more cultivation of barley (in the Levant) which is more saline resistant. So soil erosion, saline contamination, depletion of nutrients, loss of the integrity of watersheds have been inherent problems of grain production based on annuals since the beginning of grain agriculture. The solution to these problems was expansion and applied technology, often creating further problems down the road.

The cultivation of annual grains melded with human potentialities in particular ways that also pushed the system in the direction of expansion. Early agriculture would have required an expansion of the division of labor simply because the simultaneous deployment of two different strategies of survival created more complexity to execute. In simplistic terms the number of jobs necessary to the survival of the band would have expanded as they began to cultivate grains because initially the multitude of tasks associated with cultivation were additive.[18] This might seem like an insignificant matter, but in consideration of collective life, it is not. We are well served to keep in mind that in the currency of evolution small insignificant changes often become amplified. If humans hadn't had a species potential for cooperation and the ability to extend the division of labor (and they did) it would have been difficult to engage agriculture: that is, to take the first step. Culture, as we have already discussed, was an important mechanism in the deployment of the division of labor, and the division of labor had efficiency benefits that fed back into the system and pushed it along. There is no question that part of the coevolutionary match between humans and annual grains owed to the human capacity for executing and expanding a division of labor and engaging collective work.[19]

The division of labor was further extended and its inherent benefits garnered through the coevolution itself. Here the benefits of the division of labor around routinized work offered by annual grain production come to complete fruition. Annual grains became increasingly uniform and standardized in their patterns of growth and maturity through their interaction with humans (nonshattering seeds that mature at the same time, for example). So as grain characteristics were honed so were the conditions for increasingly routinized and standardized work, opening the door to efficiencies garnered therein through the division of labor. This added structure and synchronization to human society and remapped the boundaries and expression of their cooperation. The historian James Scott tells us the routines of agriculture were daily and seasonal where sowing, weeding, watering, cutting, bundling, threshing, gleaning, winnowing chaff, sieving, and drying dictated the rhythm and structure of the day. In his words: "These meticulous, demanding, interlocked, and mandatory annual and daily routines . . . strap agriculturalists to a minutely choreographed dance steps . . . they insist . . . on a certain pattern of cooperation and coordination." Again Scott claims that humans were "disciplined and subordinated to the metronome of our own crops. . . . Once *Homo sapiens* took that fateful step into agriculture, our species entered an austere monastery whose taskmaster was mostly of the genetic clockwork of a few plants." Again, and I repeat, Scott is left to the following conclusion: "I am tempted to see the late Neolithic revolution, for all its contributions to large-scale societies, as something of a deskilling"[20]

It was a deskilling, but there was efficiency in this. And perhaps more importantly the collective effort and standardization bound humans together in material life around the cycles and demands of annual grains in a completely unprecedented way. Human beings formed a collective whole around grain cultivation, a "pattern of cooperation and coordination" that deskilled the individual but gave very clear benefits to the collective in the form of efficiency of production through a more attenuated division of labor and routinization of work. There was great energy surplus in this. In fact, the structure of material life associated with grain agriculture was dictated by the needs of annual grains and the coevolution with them and bound humans together in material life, just as the diversity of plants and animals utilized by hunters and gatherers and their embedded strategy of material life around mobility dictated the structure of material life of hunters and gatherers. The knowledge and skill associated with hunting and gathering resided in the individual and in the quality of observation and understanding of a varied and complex nonhuman world. Engagement with this world was simply not amenable to standardization and rationalization in the way it came to be with agriculture. With grain agriculture humans became cogs in a larger machine, part of a larger interdependent whole. It would be hard to find a better match for employing efficiencies inherent to the division of labor, capturing economies of large-scale production, and playing on the human potential for cooperation than grain agriculture.

Cultural elaborations abounded in the largesse of surplus energy produced through grain agriculture. Technological change, built environment, and the many manifestations of institutional life all interfaced and fed into the positive feedback loops of the system. Insects build civilizations, but they don't build pyramids for their kings and queens; they don't levy taxes to support hierarchy; they don't develop market exchange to embellish, redistribute, and expand surplus; and they don't use money. And the material manifestations of an expanded social division of labor for insect cultivators do not extend beyond the needs of cultivation and reproduction of the colony. Not so for humans who develop a more elaborate division of labor around the expanded material and institutional life enabled through surplus. Thus, unique human attributes embellished the economic superorganism, creating a particularly powerful and virulent form. Kings, priests, slaves, craftsmen of many varieties, merchants, warriors, healers, and midwives emerge in the wake of the surplus system and so do pyramids, irrigation systems, plowshares, military accoutrements, villages, houses, furniture, and the list goes on.

Let me elaborate briefly on unique human capacities to enhance the system. Many species have the capacity to use tools; however, humans' capacity for observation, mimicking, problem solving, and the ability to pass on knowledge, know-how, and an expansive built environment gives them a particular aptitude for technological endowment. This capacity is cumulative and as already mentioned was parlayed into counteracting the ecological problems of grain agriculture in the short run. Clearly human ingenuity and intelligence were not absent from the complex tapestry of grain agriculture. These fed the expansive dynamic that had taken hold by enhancing grain production with innovation. But it is essential to understand that innovation was not done with an ecological mind; it was not done to counteract the structure and dynamic of the system; and it was not done using "nature as measure."[21] Rather, these were elaborations of the system. Another thread in its complex tapestry.

Intensification of the system was also augmented with the establishment of the institutions and other cultural accouterments of the surplus energy created in the system. These similarly amplified the feedback loops of the surplus system. Surplus created the possibility for hierarchy, and hierarchy became yet another manifestation of the extended social division of labor. In its own way hierarchy fortified interdependency. And hierarchy similarly demanded greater surplus because it is by definition a proportion of the population that must be supported by the work of other humans; thus it is another of the many feedback loops inherent in agricultural systems. A class structure emerged around the production and distribution of surplus, but it also fed back into the system and pushed the system along. Unfortunately, humans have a great capacity to engage hierarchy. This tendency was likely reinforced by the onerous and boring work of agriculture. We might keep in mind that if participation in agricultural life had been that rewarding it is unlikely that 80 percent of the human population would have become literally enslaved by it. And expansion and fortification of the system led to the need for military dominance, which required recruits and sufficient surplus to support them. Kings and queens, imperial armies of defense and aggression, priests, and slavery are but a few examples of the expanded social division of labor that emerged from the surplus system of grain agriculture continually increasing the demand for grain production.

Patriarchy is an example of a particular form of hierarchy. Patriarchy assured that the reproductive roles of women augmented the system—women provided the population to work the fields, the bodies for the armies of expansion and defense, the heirs to thrones and property, and the population that fortified the expanded social complexity associated with surplus. Women's lives became ever more narrowly circumscribed by reproduction, and a larger extended family offered further possibilities for expanding the division of labor.[22] Reproductive rates increased naturally as a result of sedentary life, but population growth was a foundational part of the expansive system dynamic—more grain production, more population, greater division of labor. And the attrition from lethal crowding diseases necessitated a vigilance and attention to reproduction.[23] Higher reproductive rates became part of the new ecology of annual grain production and a participant in the expansionary spiral. The knowledge and practice of low birth rates practiced by hunters and gatherers were replaced with a dynamic of high birth rates that reinforced the system.

Hierarchy serves to justify who works to produce surplus and who lays claim to the surplus, and it becomes foundational to the social relations of agricultural production. Hierarchy is an elaboration of the social division of labor that expands with agriculture and tensions in the system will surface here, especially if problems with continual expansion and surplus production are threatened. But hierarchy is only one form of the social division of labor associated with agricultural surplus. Agricultural surplus allows for the production of a greater variety of things (vessels for storage, military equipment, built structures, etc.) that are now possible with surplus energy (and necessary) around sedentary life. In addition, elaborate forms of homage to gods and their earthly representatives emerge and are found in kings, queens, and priests.[24] And there are efficiencies inherent in these specializations as well. One can be a potter or a weaver or a priest for that matter and get very good at it. The social division of labor expands and greater interdependence in material, spiritual, and hierarchical life along with it. And with the rise in population and expansion of the social division of labor greater bureaucratic dimensions of cooperative life are necessary. Complexity takes form and demands ever more energy as Joseph Tainter rightly argues.[25]

An expanded collective human enterprise became dependent on the foundation of the agricultural system and its potential for surplus. It took only six thousand years of the engagement with agriculture for vast city states to emerge with their elaborate hierarchies, armies of defense, and expanded material and bureaucratic life. While the interdependent structure and the dynamic of expansion are universal attributes of the agricultural system there are clearly institutional, technological, and cultural augmentations to the system that are particular to humans. The extensive interdependence in material procurement and its cultural and institutional stanchions push the human propensity for cooperation in an unsavory direction—that is to say, in the direction of loss of individual autonomy—and humans becoming cogs in the economic machine (superorganism). Humans are bound together in material life in this system with its profound interdependencies. To escape this system literally meant endangering your material survival (and, incidentally, still does).

Perhaps the most powerful of the institutional embellishments of the agricultural system are the elaborations of its economic institutions. Markets, expanded networks of trade, debt, money, taxes, property rights, and class are the most obvious. They not only served to distribute the surplus but also created an institutional fabric around surplus that itself fed back into the system. In the end these economic institutions take on a life of their own—further adding to the expansionary and self-referential dynamic of the system. Some ten thousand years after the agricultural system began the institutional and cultural embellishments of surplus take on their most pronounced form with the rise of capitalism with its particular architecture of class, private property, profit, and the technological embellishments that facilitate its maturation. Capitalism married to the industrial revolution exaggerated the duality between humans and the more-than-human world that began with grain agriculture. That duality reaches an almost frenzied pitch as human economic life is temporarily relieved of any sense of limits through the use of fossil fuel. The institutional life of surplus now manifest in the accumulation of capital moves rapidly forward and basic biophysical connection between economy and Earth is lost. No one really understands how to disengage the present rendition of the economic superorganism. I will elaborate on the evolution of the economic institutions of surplus, and in particular capitalism, in section 3. Note that markets begin as institutions of redistribution in a world of surplus, but over millennia they fold back on themselves to create a powerful augmentation to expansion and surplus in the form of a profit system. A more institutionally elaborate variant of the agricultural system emerges.

Despite the many problems associated with agriculture around annual grains this mode of production went forward with great acceleration and embellishment and came to dominate the material life and evolutionary arc of human beings. In a brief time after the first seeds of domesticated annuals popped through the cultivated soils of the Holocene vast state societies had emerged. The story was the same no matter the place, or time, and no matter the annual grain.[26] Humans had wandered far from the embedded existence of hunting and gathering and its basis in human observation, minimalism, mobility, and intelligence. They now took on the collective expression of an economic superorganism, and *Homo sapiens sapiens* became the energy maximizing, interdependent, and expansionary species *Homo sapiens agriculturii*.

Humans did not walk through the door of agriculture simply out of intentionality, inventiveness, and cultural proclivity in the constant struggle for food. In fact, it is easy to argue that the struggle for existence after agriculture became more formidable than it had been in the several hundred thousand years of human history that predates that momentous change. A distinct material system was established with grain agriculture and, in a sense, humans have merely been along for the ride ever since. Despite the human capacity for adaptation through culture broadly construed (embodied in institutions, beliefs, and technology), it would appear that these have mostly been directed toward embellishing that system. We humans arrive at the gates of the Anthropocene with hubris and misguided assessments of what has happened to us, how our relationship with Earth has been altered, and how to change our trajectory. We appear to have arrived with an inadequate appreciation for the system dynamic that carries the day and the way our particular proclivities feed into rather than counter it.

Chapter Five

"A Species Out of Context"

I have argued that agriculture was a problematic turning point for humans. Humans arrived at the doors of the Holocene as one thing (*Homo sapiens sapiens*) and left as an economic superorganism (*Homo sapiens agriculturii*). There is much currently written about the prospects (and fears) surrounding the possibility of the genetic engineering of humans.[1] Yet on closer reading of human history we can conclude that humans can be radically altered without changing their DNA. The potentiality of humans can be expressed in dramatically different ways as the evolution from *Homo sapiens sapiens* to *Homo sapiens agriculturii* tells us. And the formation of context is a complex unfolding that ultimately defines the relationship of humans to each other and to the more-than-human world.

Organized through the agricultural system, humans were no longer in community with Earth in the way they had been as hunters and gatherers. Instead they were "a species out of context."[2] That is to say the context that had chiseled the human species into *Homo sapiens* and defined species existence up to that point. Humans were no longer a minimalist, slow-growing species embedded in the rhythm and dynamic of the more-than-human world in economic life, where the more-than-human world carried on its self-willed otherness and humans adapted around it. Instead humans became a structurally interdependent species pulled into the vortex of the feedback loops of an expansionary, interdependent agricultural system, a new whole to material life that formed a duality between human material life and Earth that had not been there before.

The ensuing ten thousand years of the unfolding of this system has now reached an apogee with global capitalism. Capitalism is staggering in its interdependencies, its divisions of labor, its relentless expansionary dynamic, and its reach into every nook and cranny where surplus, now in the form of profit, can be extracted. The system reaches into the depths of the human heart and the genomic endowment of the planet. Nothing is sacred. The duality between humans and the more-than-human world and the expansionary dynamic that began with the agricultural revolution has taken on the ring of a death knell. Earth has not only become resource, capital, ecosystem services, depository of infinite externalities (pick your moniker), but the human domination of Earth has reached a terminal stage. We humans are exterminating the world's species and undermining ecological integrities and natural cycles that we rely upon to support the almost eight billion of us. It is a gross understatement to say we are no longer in community with Earth. In fact, we are being carried along by a system that we appear unable to alter, and most of us function as cogs in its machinery at best, or are superfluous to the system at worst.

If there is a silver lining in this story it is this—the inclination toward the economic superorganism has not been a comfortable outcome for humans. It has distorted the intimate relationship humans share with Earth—an intimacy so fundamental that it provides the basis of biological existence, the seeds of human creativity, the power to calm the human psyche, and some would say the power to direct human ontogenesis toward healthy maturation.[3] This has not gone unnoticed. By now there have been millions of pages written about the environmental crisis. And the more poetic writing of the transcendentalists, and the Wendell Berrys of our time, speak to the heart and soul of the matter. Berry tells us what to do when despair grows in us: "Lie down where the wood drake rests in his beauty on the water" and "come into the peace of wild things." He strikes a chord that cannot be overlooked.[4] Nature deficit disorder has entered the lexicon and reality of our time. The move in the direction of the economic superorganism also has been uncomfortable because it has fomented alienation, perverted our sociality and propensity for cooperation, and has opened the door to hierarchy and exploitation. It has made each person a bit player in a powerful system over which each feels incapable of exerting control or altering in any significant way.

Finally, the inclination toward the economic superorganism has been uncomfortable because it has heightened the probability for system collapse. Culture and inventiveness aside, the wedge between humans and Earth has been driven ever wider with each passing millennium and now—in the face of the upper neck of the exponential curves that define the human/

Earth interface—each passing decade. The human economic superorganism becomes more irrevocably self-referential as humans remain unable to extricate themselves from the present manifestation of the agricultural turn, a massive global system oscillating between growth and stagnation in the face of ecological collapse. We live in a state of collective despair over collapsing ecosystems, the mass extinction of the other species that share the Earth with us, and the warming of the planet by our activity—all the while we tout the benefit and necessity of economic growth and map out our strategies for personal success in what is an unsustainable and seemingly unassailable system. It is contradiction and confusion that define our historical moment. (I will return to this in part 3.)

Out of our Pleistocene potential humans became an economic superorganism and have been moving along this path ever since. Despite the adaptability offered by culture, we seem to be engaged in a trajectory we are unable to alter.[5] Culture is a quick adaptation to changing conditions by evolutionary standards, but it is slow and ineffective in the face of the upper neck of exponential curves. And culture, with its institutional and technological trappings, appears to have reinforced and augmented the structure and dynamic of the economic superorganism and not run counter to it. The line between culture as a mechanism of change and culture as an embellisher of the system is murky at best. It is reasonable and important to ask in the presence of this complexity what it means to be human. Is it not as human to be *Homo sapiens agriculturii* as it is to be *Homo sapiens sapiens*? In some sense they are both expressions of human potentiality in the endless play of evolution, but suffice it to say they are not expressions without their unique consequences—and it pays to have some sense for that. Stephen Jay Gould tells us that "humans are animals and everything we do lies within our biological potential. . . . But . . . the statement that humans are animals does not imply that our specific patterns of behavior and social arrangements are in any way directly determined by our genes. Potentiality and determination are different concepts."[6] But potentiality does take form, and when it does the outcome may be as determinant of our evolutionary path as would a change in our genome.

Although I veer away from claiming one human expression is natural and another is not, what I will claim is that without countervailing interventions we will continue on the path we took with the Holocene—the path of the economic superorganism. We will finish the sixth mass extinction, heighten the probability of collapse, continue the project of relegating humans to cogs in a "machine" at best (or redundancies in the economic system at worse), and continue the thwarting of the nuanced balance between individuation and belonging and the way it rears its ugly head in our present tendencies toward human supremacy, personal greed, and the various manifestations of maturation gone awry (e.g., fascism and nationalism). Humans will take a final and irrevocable step to a different world where it is no longer possible to reside in an environment of rich and numinous otherness. This outcome is the legacy of the context created by the agricultural revolution and the engagement of the agricultural system. Human beings will be forced to abandon any pretense of an intimate interchange with Earth—an interchange that programmed their ontogeny, captured their imagination, honed their capacity for observation, created their scientific minds and helped to chisel the human as an intelligent, minimalist, big-brained primate that lived in community with the Earth for most of human history.[7] Human beings will be forced to abandon the solace and sounding board of the more-than-human world to moderate and soothe the anxiety of being human. They will be stuck with the simplified, brittle ecological landscape of global capitalism and its particular and perverse expression of human cooperation, not to mention the duality between humans and Earth that is embedded in its fabric. I suggest that at the eleventh hour this outcome is worth a second thought. The consequences of the agricultural turn are clearly not captured by the accepted narrative that agriculture was a beacon along the road of progress.

The truth is that we are still as much Pleistocene as Holocene. Were you to drop a human baby into a hunting and gathering band of the late Pleistocene he/she/it would hardly be a fish out of water. So it is important to explore more fully the difference in these two ways of living on Earth as we take an irrevocable turn. For millennia following the onset of agriculture the door was still open to the possibility of touching the Pleistocene. We could be reminded of it, at least around the edges of our daily lives. Metaphorically (if not literally) speaking there were wolves at the gates of civilization, and the ecological integrity and rich otherness of the Earth was intact over much of it.[8] But the trajectory of the agricultural system has matured, and the economic superorganism is enveloping Earth. The existential threat is not merely whether human beings will survive climate change but rather what they become as the economic superorganism reaches its apogee. This is the existential threat that humans fail to perceive in all the hand-wringing surrounding our present circumstances. The threat is that we are becoming irrevocably *Homo sapiens agriculturii*, members of the economic superorganism. There is no solution to this trajectory unless we can alter and contain the economic system and downsize the human presence on Earth. Herein lies our problem: the basic impulse of the system is not in the direction of downsizing but the opposite, and it isn't clear how much power we have to alter its course especially given the capacity of culture to embellish the system.

In light of this challenge, it pays to go back to the economic system of the Pleistocene so that we might think more expansively about where we stand. Here we discern the rough outline of a different context of human life with different consequences. Many years ago, the anthropologist Stanley Diamond said, "The search for the primitive is the attempt to define a primary human potential. Without such a model . . . it becomes increasingly difficult to evaluate or understand our contemporary pathology and possibilities."[9] So I go back to the hunting and gathering past because this mode of production was the foundational expression of human potentiality out of which we emerged as *Homo sapiens*—it represents something worthy of reflection. Clearly we became fully human around this mode of production. This comparison provides an important point of reference to think about the contextual expression of human existence and the consequences of taking the agricultural turn.[10]

Hunting and gathering must be viewed as a particular type of material system, a particular "mode of the production," forming a different context to human material life. It was a system that nurtured and gave rise to *Homo sapiens* as an observant member of an Earth community: a slow breeding, big-brained, resourceful, minimalist species. I describe a generalized picture of the hunter-gatherer just as my description of the agricultural system is a generalized description of that system. Neither are meant to capture historical detail but rather to interpret a broad stroke of history. And I realize that I enter into controversial debates about human prehistory. Perhaps the most contentious is my assertion that humans lived an "ecological" existence, embedded in Earth's community before the engagement of agriculture and that this is one of the overarching characteristics of this system and of the lived experience of human life for much of our history. This runs counter to the popular narrative that humans were responsible for the extinction of Pleistocene megafauna, for example; that they developed technology by virtue of their big brain that allowed them to dominate; and that they were a warring, aggressive, and expansionary species struggling for survival long before the agricultural revolution. I emphasize a different interpretation of human prehistory. I am not suggesting that our preagricultural ancestors lived entirely in ecological balance, nor that scarcity or conflict were entirely absent. I am merely suggesting that the hunter-gatherer was not engaged in a *system* that assured or engendered these outcomes.

The hunter-gatherer lived in a world constrained by the need for mobility that helped to define the structure and dynamic of species life. Marshall Sahlins clarified this for us.[11] And while hunter-gatherers needed energy, they weren't involved in a system of net energy expansion. Again, this runs counter to the popular narrative that prior to agriculture human life was a daily struggle for survival and the preoccupation of humans was continual striving to obtain more energy.[12] They certainly needed enough energy and were bound by an overall energy return on energy invested (EROEI) mandate; that is, that the energy they expended in aggregate had to be at least equivalent to the energy attained.[13] Yet before the agricultural revolution humans were not structurally and dynamically geared toward expanding their energy budget, and scarcity was for the most part not their primary problem. As Sahlins tells us, the hunter-gatherers were the original affluent societies—their "wants are finite and few, and technical means . . . adequate."[14] In other words the system of hunting and gathering was not a high-energy approach to the perennial problem of entropy. To the extent that the hunter-gatherer managed to increase the efficiency by which they attained energy the benefit was taken in leisure, in ritual, in play—not in endlessly feeding an expansionary dynamic. In other words, obtaining energy did not become autocatalytic. It did not create a positive feedback loop of expansion.

The structure of this mode of subsistence was a matter of mobility and was punctuated with an embodied ecological knowledge residing in each individual—obtained through observation, mimicking, advanced social learning—and passed down through generations in a cumulative fashion. This is what it means to be smart and cultural. The many modalities of human existence (ritual, belief, social, economic) were interconnected. And although the hunter-gatherer lived in the community of small bands that sometimes amassed temporarily into larger groups, each individual human had great capacity to sustain themselves independently. Specialization did exist, but each person had a breadth of knowledge and proficiency in the provisioning of material life.[15] Material interdependence in this sense was minimal.

Material independence did not mean there was no cooperation in material life. Women depended on women for child birthing and rearing of children and gathered together; men (and some women) worked cooperatively, especially around big game hunting. Once they downed a large animal there was sharing. And there was, of course, a sexual division of labor mostly connected to biological differences. In so far as defense was necessary the clan was a cohesive whole, and sticking together was a good strategy for defense when and if it was needed.[16] But despite these expressions of cooperation humans were mostly independent in the provisioning of daily subsistence and were not mechanistically connected in their daily material provisioning with autocatalytic feedback loops and an endless expansionary dynamic. As Richard Lee observed, there was "a degree of freedom unheard of in more hierarchical societies. In the organization of production foragers could work their own schedules. They did not experience the 'benefits' of work discipline and regulation imposed by centrally organized agrarian states."[17] They were not bound by the "metronome of a few crops."[18] These differences were the outcome of a different material system where humans didn't sweat and toil in synchronized fashion from dawn until dusk to cultivate annual grains.

It does appear to be "fact" that *Homo sapiens* radiated out of Africa at least a hundred thousand years ago and that before doing so they had attained significant cultural and technological proficiency. These were expressed in "hunting, fishing, long-distance travel, clothing, hearths, fire management, symbolic thinking, art, decoration, social cohesion."[19] There is no question that culture broadly construed and intelligence (both honed during the Pleistocene likely in response to the dramatic climatic changes that occurred throughout this geological period) gave the advantage of quick adaptation that would have otherwise been absent. The survival strategy of the hunter-gatherer was as much the deployment of a big brain manifest in ecological knowledge and reading the more-than-human world as it was brawn and domination. With larger brains humans evolved into a resourceful and observant primate with the capacity to engage in robust social learning, with the ability to strategize, categorize, and to read the ecologies (and their processes) through which they moved. These became integral to flourishing. Each individual participated in subsistence activity endowed with these evolved capacities for knowledge and their extension through culture, which surely involved technological development. However, that development was done in context, and that context was not scarcity: it was, to use the vernacular of Eileen Crist, "an abundant earth."[20] While there is little question that humans were smart and had attained culture broadly construed by the time they began to migrate out of Africa, what is up for debate is why they migrated.

One narrative is that the diaspora occurred because humans depleted local environments primarily through domination elicited through innovative hunting technology. In other words, in their fight for survival, their big brain got them in trouble.[21] The story is that smart, resourceful humans in the struggle for survival innovated to the point of being able to overshoot ecological balance. It is a fact that *Homo sapiens* began to hunt big game and that they developed more sophisticated technology to do so by the time they began to migrate out of Africa. It is an interpretation of this fact that the radiation itself was caused by population pressure and ecological stress brought about by human capacity to dominate through their developed technologies in an effort to expand access to energy; that is, they had become un-ecological. There is another story to be told, but we will never know which story is true because prehistory is mostly a matter of speculation and storytelling based on limited physical evidence. Here preconceptions honed in the present can be magnified in their influence in the face of limited evidence.

There is little doubt that hunting and its technological development made for good strategy for enhancing the survival of human beings. The hunt itself was perfected by many human attributes. Hunting was mostly about cooperation and observation; that is, observation of the ecology of the animals that were hunted and cooperation among hunters.[22] Individuals, for the most part, did not take down big game alone. And hunting came to involve a great deal of ritual—one of many examples of the interconnectedness and integration of modalities of existence. Although it was one part of the many strategies of survival for humans, the dependency on big game for daily sustenance is questionable. It certainly varied from clan to clan, but it is fair to say that big game hunting was mostly not a central axis of food provisioning. Gerta Lerner points out that "in most of these [hunter and gatherer] societies, big-game hunting is an auxiliary pursuit, while the main food supply is provided by gathering activities and small-game hunting, which women and children do."[23] The point is not that the meat provided through large game hunting associated with men was unimportant but simply that it was much less central to survival than is commonly thought. It should not be the central axis of our interpretation of human prehistory.

Women and children roamed together gathering and snaring small animals, and this provided the greater part of their daily food needs. Women with young children and young children themselves did not participate in big game hunting because it involved both extensive running at times and not being seen nor heard at other times. Contrary to popular belief women and children did not wait half starved for the hunter to return home to feed them: rather, they were geared to fend for themselves. And boys came to understand the full spectrum of plants and animals available to them and developed the power of observation necessary to obtain them: this was done by engaging in gathering activities and snaring small animals with their mothers/aunts/grandmothers over years of maturation. In the end, big game hunting was likely as much ritual and initiation as it was sustenance. There is no doubt that hunting technologies and the inventiveness of humans inserted into long-standing ecologies might have created havoc and imbalance from time to time, especially when connected with other changes like the retreat of the glaciers some twelve thousand years ago.[24] But the thrust of the hunting and gathering system was not in the direction of overkill and ecological overshoot. Almost everything about the hunter-gatherer and their system led in a different direction.

The remarkable thing about hunting and gathering was not how expansionary and unecological it was but how contained, minimalistic, and embedded it was. It would be just as easy to emphasize hunting and gathering in light of its imperative to maintain mobility, low birth rates, little expansion, and ecological knowledge as it is to claim that smart *Homo sapiens* with a penchant for violence, domination, and in the struggle against entropy expanded and destroyed ecological balance in some ceaseless drive pitting humans against each other and other species and exterminating the latter. The latter story is emphasized because it appears to fit the pattern of our present system, a system that does engender a preoccupation with intelligence manifest in technology and a tendency toward ecological destruction in a dynamic of expansion—a system preoccupied as it were with scarcity. If there is an unecological bent in us, it took hold with the engagement of the agricultural system. The effect of the knowledge and inventiveness of the hunter-gatherer would not have been always perfectly calibrated to local ecologies, but the hunter-gatherer clearly was engaged in a mode of production, a system, that was inherently minimalist, mobile, low in energy demands, and mostly embedded in the rhythm and dynamic of the ecologies through which they roamed. The material strategy of the hunter-gatherer was structurally configured to move in a different direction than we now move.

I am compelled to add one more assertion with regard to human intelligence and knowledge and its role in the hunting and gathering mode of production. Of the many strategies for survival employed by the hunter-gatherer, population control was paramount since the line between stability and overshoot was certainly quite fine. It is well understood that mobility itself affected birth rates, but it seems likely that maintaining low pregnancy rates and by extension birth rates must also have been an intentional strategy based on knowledge and not just an unintended result of constantly moving. One can argue that the knowledge of controlling pregnancy was easily as important as the knowledge of hunting to human survival, yet this "technology" left no artifacts—it was not chiseled out of stone, and therefore it is given a sidebar in human prehistory. Logic tells us that the practice of infanticide would likely have been a last resort in controlling population. It was the knowledge of preventing pregnancy and aborting unwanted pregnancies that were important and that involved extensive knowledge about female bodies and Earth. An expansive knowledge of plants and the cyclical nature of ovulation would have allowed them medicinal aptitude around birth control, abortion, as well as birthing. I think we understand very little about this technological knowledge because once again it leaves no artifacts and lies outside the narratives we tend to emphasize.

In summary, hunter-gatherers were loosely structured in feedback loops around mobility, low birth rates, observation, and yes, inventiveness and cooperation. This world chiseled humans as intelligent beings with a nuanced sense of the relationship of the individual to the whole—to the whole Earth—and to the community of humans with which they were affiliated. The more-than-human world provided the context of human life not simply as material resource but as an organic whole of existence. There is no doubt that inventiveness altered their world, but the imperative of that inventiveness was not expansion, surplus, and domination. And let's face it, it's difficult to kill a big animal with your bare hands, and it's easier to dig up tubers with a digging stick. Inventiveness need not imply an endless push for more energy in the face of scarcity, nor need it imply an inherent penchant for dominion: rather, it was simply a matter of the strategic resourcefulness of an industrious, cooperative, and observant primate embedded in a rich and inspiring more-than-human world.

This expression of human life must be juxtaposed with that of *Homo sapiens agriculturii* who were different in almost every way. The comparison helps us to appreciate how contextual human beings are. Agricultural humans were structurally and mechanistically interdependent in material life. They could not sustain themselves independently. They were net energy maximizers, involved in complex feedback loops between population and its structure in the division of labor, institutional fabric, and energy that pushed the expansion of grain production. They were not nurtured with the same keen and nuanced ecological awareness of the hunter-gatherer and in fact were engaged in a system that degraded ecologies, cordoned off the human enterprise, and placed the more-than-human world in a simple dichotomous key: us and them. Human inventiveness was in service to net energy expansion around grain production and not to nuanced ecological knowledge and embeddedness. Rather than mobility they had the necessity of staying put where fixed-field farming, urban enclaves, and possessions could expand and a self-referential bent to human life could fully flourish. This was a distinctively different expression of human potentiality, a completely different system of material life. The agricultural system established itself as a dynamic system dependent on self-reinforcing feedback loops inclined to "overlook" ecological constraints pushing the system toward collapse. Humans have been carried along by the structure and dynamic of the agricultural system in one form or another for ten millennia, and clearly it has become a more entrenched and dangerous system over that time. (I discuss the contemporary legacy of the agricultural system in section 3.)

The expansion of the human enterprise (the economic superorganism) as a result of the engagement of the agricultural system has been astounding, especially considering that humans lived in a mostly nonexpanding system for several hundred thousand years as anatomically modern humans before agriculture began. In fact, so expansionary is the system that six thousand years after the onset of agriculture we see the rise of state societies, and by the time we hit the eight-thousand-year mark human population had increased from four million to 250 million, and this despite the prevalence of the diseases of domestication and their toll on population.[25] This is a universal dynamic, and it happens for humans around different annual grains in different parts of the world independently; yet the results are essentially the same.[26] And as the complexity of civilization increases with an expansion of hierarchy—the social division of labor and the elaboration of bureaucracy to manage complexity—the energy it took to keep civilization going increased, adding to the expansive imperative of the system.[27]

Embedded in the altered system that took hold with agriculture we find its most important consequences. The consequences manifested themselves at multiple levels of human existence. The day to day of material life became a more focused, routinized, rationalized, and integrated collective enterprise dictated by the demands of annual grains reducing the complexity and embeddedness and knowledge in the provisioning of food characteristic of the hunter-gatherer to the routines of "fixed-field farming" of cereal grains, and changing the boundaries of cooperation. James Scott captures this beautifully when he tells us, "Farmers, especially fixed-field, cereal-grain farmers, are largely confined to a single food web, and their routines are geared to its particular tempo." And in this there is a "substantial narrowing of focus and simplification of tasks," but perhaps as important are "growing chains of dependence."[28] Grain agriculture did not nurture an awareness of multiplicity and complexity, nor did it nurture the imaginative human mind except in those that could escape its day-to-day drudgery by living off the surplus produced by others. The expansive knowledge of the hunter-gatherer and their penchant for observation were reduced to the demands of a few annual grains. The participation in material life was less interesting, less imaginative, and more interdependent in a machine-like way. And in a sense, it was perversion of both human interaction with the more-than-human world as well as the human propensity for cooperation. Human material life was encapsulated around the vortex of grain production. The expansion of the social division of labor and the institutional and built manifestation around surplus further elaborated interdependence around grain production, and human material life became ever more a thing unto itself, cordoned off from wider ecologies ultimately concentrating human life in urban centers around a built human environment. Urban enclaves with walls for protection from invaders stand as a metaphor for the walling off of the human enterprise from a fluid interchange with the more-than-human world—but it was also a physical reality.

The quality of the daily interaction of humans with the Earth had been irrevocably altered, making what was once communion and ritual, imagination and observation, instead drudgery and a relationship of resource, pest, and predator. The philosopher Bill Vitek describes what agriculture did to the human understanding of the more-than-human world: "Ancient and indigenous understandings of a wild, creative and sacred Earth were interrupted, and driven underground, and nearly eliminated by the slow development of annual-disturbance grain agriculture 10,000–12,000 years ago. With its powerful dualisms pitting crop against weed and livestock against predator, agriculture established attitudes that nature was to be subdued or ignored."[29] Humans responded to the challenges of agriculture with ever greater innovation, not with greater ecological awareness. They dug in so to speak.

Duality also came to rest in altered worldviews that emerged after agriculture took hold. Think of the culmination of a changing worldview embodied in the monotheistic religions. Bill Vitek describes the transformation in beliefs: "India, China, the Middle East, and the Mediterranean basin all saw transformational transitions away from tribal, animistic, and polytheistic cultures with gods, goddesses, and rituals connecting humans to nature . . . The Axial Age, in other words, reordered the metaphysical landscape and gave us the three primary and distinct metaphysical units we live with today: God, humans, and nature. God is up there, all powerful and loving; nature is down here, mostly passive and resourceful. Humans are in between, God becomes separate; nature becomes secondary; and humans share both divinity (our reason, mind, or soul) *and* nature (our bodily desires and appetites). We live in a world divided by a Jewish-Christian-Islamic God and a Greek and later Enlightenment rationality—a world made separate and very much unequal"[30] This alteration in worldview came directly and unequivocally out of a duality created by the agricultural system and ultimately came to be embodied in Enlightenment thinking.[31] Finally the alteration in worldview also came to rest in our present belief in "the market," which as far as I can tell is replacing the belief in a monotheistic God.

In the structure and dynamic of the agricultural system and its separation from ecological grounding is written the repeated history of collapse. The system dynamic is powerful, but eventually it becomes overpowered by ecological degradation and the tensions (class) within the system. Short-run technological and institutional solutions can add to the time until collapse (irrigation systems, renewable energy, institutions to redistribute surplus), but a system of surplus eventually comes up against its antiecological bias and its tendency to function separate from ecological grounding. Agricultural systems meld technology and institutional life and take advantage of efficiencies to expand and mitigate against external limits until they can't. The rise and fall of civilizations are as reliable as the sun rising and setting. The story is consistent and predictable. In the Far East, in the Near East, in the Americas: in all places the stories of collapse abound. The economic superorganism is particularly dangerous for humans who have the capacity to embellish the system dynamic with cultural life broadly construed, a penchant for inventiveness, and the force of ideology to convince them of things that aren't true. The multiple feedback mechanisms of the economic superorganism for humans function without much reference to the broader ecologies unless they are forced to do so, and usually the response is to find a way out of a particular ecological bottleneck.

Over the millennia and since the agricultural revolution the wedge of duality between human economic life and Earth has been driven deeper. Human material order and its attendant belief systems and institutional arrangements have become more insular and more removed from embedded ecological knowledge. Humans imagine "redemption" from the ecological consequences of their path through their ingenuity, institutional arrangements, technological abilities, and their capacity to work together to problem solve. Yet, quite reliably, human ingenuity and institutional life forge a greater duality and further cement the self-referential economic superorganism. As we have seen, these are mostly accommodating to the system. And these accommodations can become quite elaborate as the ultimate development of the market economy and its ideology demonstrate. Though humans can never be biophysically disconnected from Earth, in some very real sense they function in an economic system that stands apart from the Earth. We might adopt a more critical perspective to think about the ability of humans to work together to problem solve. As this inquiry demonstrates, human sociality is a complex matter in unfolding system formation.

There is no better way to understand the amplification of the agricultural system than to highlight the institutional stanchions of surplus otherwise known as economic institutions. These eventually include money, debt, markets, taxes, trade, embellishments of private property, justifications of class relations, and ultimately the ideology of "the market." Capitalism is a system that functions as if it is wholly disconnected from the Earth especially once it becomes tied to fossil fuel. It is an expression (in institutional form) of the duality that began with agriculture—in fact it is so profound that as we stand on the brink of disaster we mostly refuse to acknowledge that there is any problem with this institutional interpretation of surplus. We still imagine redemption in terms of the market system and its ability to channel our technology and human ingenuity.

Part III

Apogee

Chapter Six

Capitalism: A System within a System

There are many modern-day prophets of the environmental crisis who offer words of hope. Unfortunately, words of hope are often couched in a poor understanding of *the economy* and the prospects for changing it. I am sympathetic to this inclination because confronting the bone and sinew of the economic system only seems to reinforce a sense of hopelessness in the face of this war between economy and the Earth. But I am not trading in the currency of hope; I am trading in the currency of understanding because this is what is needed to know where we stand. Greater understanding can only take us to a different level of engagement with our circumstances. There is no choice but to wade into the nitty gritty of our current economic system even if the risk is to take the wind out of the sails of optimism by doing so.

Begin with what appears clear. All the characteristics of the economic superorganism are exaggerated with capitalism: its expansionary dynamic and production of surplus, the duality it creates between humans and the more-than-human world, the material interdependence and autocatalytic feedback loops of the system. In this sense it is appropriate to see capitalism as a system within a system. We remain on the trajectory established when the agricultural system took hold. Yet the question remains: What exactly is this system within a system? How is it unique in its expression of the economic superorganism, and how is it different? Our understanding of it must be clear enough to decipher what it is we're trying to change in order to realign humans with Earth. Are we trying to make a more humane and loosely sustainable capitalism or repair the juncture between humans and the more-than-human world that began when the agricultural system took hold? These are not the same thing. Capitalism can be made better or worse; there is no question about that. But making it better or worse is not the same as halting the trajectory of the system and mending the duality that now exists.

In previous chapters I have presented a deeply materialist interpretation of the formation of the economic superorganism, one that reaches into the fabric of evolution and the formation of a universal system that crosses species. The fact that it is an order shared by other species means that in meaningful ways humans are not unique but instead part of universal processes and system dynamics that govern all of life. Agriculture is a powerful and universal system, and its formation is not simply a story of human ingenuity tapping productive forces to confront parsimonious nature. Nor is it simply a story about growing food. It is also the story of the formation of a different material system, a new economic whole around cultivation—in the case of humans, the cultivation of annual grains. The etiology of the economic superorganism is a complicated and iterative dialectic of system formation made all the more complicated for humans by their penchant for culture and their intelligence.

Once a powerful agricultural system emerged it had (and still has) its own power, and history tells us that unique human attributes and levers of change have been reliably accommodating. The economic structure and dynamic that formed around grain agriculture has held true to course in the ensuing ten thousand millennia. Simon Lewis and Mark Maslin point out that there was an ongoing process of change after the agricultural revolution but that the change orbited around the magnetic force of the agricultural system. In their words: "People often think that in the thousands of years following the rise of agriculture human societies were static. They were not. . . . People innovated. They devised practical tools. . . . There were shifts from tools made of stone, to bronze and then later to iron. Food producers became more efficient as more sophisticated farming systems developed. . . . Yet little of this innovation altered what the majority of humanity did, day in, day out, in the way that the agricultural revolution had done."[1] A distinct economic whole defining the day to day of human life emerged with the agricultural revolution.

The force and resilience of the economic superorganism have now settled into its present form: global capitalism. Capitalism is a dramatic elaboration of the economic superorganism. If we are to understand it expansively, we are compelled to see it first as the legacy of the agricultural revolution and second as a system in its own right. Capitalism changed the form of surplus and expansion but not the fact of their existence; it altered human-to-human relationships, but it did not change the fact of enhanced material interdependence (nor the presence of hierarchy). Finally, capitalism drove the wedge of duality between humans and the more-than-human world ever deeper, but it did not create that duality.

To sort out the particular essence of capitalism we must begin by noting that it was fully established before the industrial revolution became powered by fossil fuel. Adam Smith's tome on capitalism, *The Wealth of Nations*, was essentially written in this world.[2] The *invisible hand*, that powerful metaphor coined by Smith, was conceived in a world mostly devoid of fossil fuel use. Thorstein Veblen captures something of Smith's world and the fact that he stood on the cusp of this change when he states, "Adam Smith spoke the language of what was to him the historical present, that is to say the recent past of his time. . . . But in the historical sequence of things he stood at the critical point of transition to a new order. . . . What had gone before was the era of handicraft and the petty trade, habitual outlook of which had become (second) nature to the thoughtful men of that time; what has followed after is the era of the machine industry and business enterprise."[3] In this passage Veblen illuminates the changing nature of capitalism and identifies its earlier form—from petty commodity production (a butcher and baker on every street corner) to fully developed industrial capitalism and the rise of the corporation.

There is no question the industrial revolution around the use of fossil fuel altered capitalism in the direction identified by Veblen, but capitalism had been fueled initially with the energy provided by agriculture, wind, forests, and falling water. Fossil fuel was not part of the mix. In its early form we find the underlying fabric of the capitalist system with its unique forms of surplus and expansion, its interdependent threads, and its particular structured duality between humans and the more-than-human world. I agree with Jason Moore's claims that "the history of capitalism cannot be reduced to the burning of fossil fuel in England."[4] And Andreas Malm is very clear that it was the unique access to falling water in Britain that fueled textile production, an industry that was exemplary in the early chiseling of capitalism.[5]

In order to understand the essence of capitalism it is helpful to look at its proximate formation. Capitalism itself distilled out of its immediate antecedent, feudal Europe, just as feudal Europe had distilled out of the decline of the Roman Empire.[6] We should understand the movement to capitalism as a gradual process of change over centuries. Robert Heilbronner captures this when he tells us of the transition from feudalism to capitalism:

"There was no single massive cause. The new way of life grew inside the old, like a butterfly inside a chrysalis, and when the stir of life was strong enough it burst the old structure asunder."[7] Over centuries capitalism grew out of the decay of feudalism—the paternalist agrarian medieval society of custom and tradition that immediately preceded it. My purpose here is not to provide the minutia of historical detail in the transition from feudalism to capitalism nor to enter into the detailed debates among the warring factions about the rise of capitalism.[8] Rather it is simply to provide a story of the emergence of capitalism that discerns it as a distinct expression of the economic superorganism.

A note of clarification is necessary. The presence of markets and trade do not in and of themselves constitute a market economy (aka capitalism), but there is no question they are a preamble to it. I would go further and emphasize that unless this distinction is understood, the particular nature of capitalism remains confusing. It is only when all of the decisions of economic life (what is produced, how it is produced, who gets what) are mediated through the market and all of economic life is brought under the dictates of the market system that we have market capitalism. As Karl Polanyi tells it: "A market economy is an economic system controlled, regulated, and directed by markets alone; order in production and distribution of goods is entrusted to this self-regulating mechanism. . . . Self-regulation implies that . . . there are markets for all elements of industry, not only for goods (always including services) but also for labor, land, and money, their prices being called respectively commodity prices, wages, rent, and interest."[9] Previous chapters have been clear: markets and trade existed long before capitalism and grew out of agricultural surplus. They are a particular institutional elaboration of surplus—one of the feedback loops of the agricultural system. Yet markets did not become the central organizing principle of economic life until the rise of capitalism in Europe. They were instead embedded in and controlled by other institutions governing material and social life; in other words, they were derivative and ancillary, certainly not central.

I take the example of feudal Europe to illustrate this point. The material reproduction of feudal life was not organized around trade and markets however present these were around its fringes.[10] Material life revolved primarily around agricultural production on the manor and the surplus it supported, and society was organized through custom and tradition and authority granted through "divine right." Surplus found expression in hierarchy, the building of feudal estates, the church, armies, knights, the guilds to produce armor, and the many craft enterprises that supported feudal life.[11]

Yet even in the seemingly staid life of feudal society change was afoot. For example, the transition from the two- to three-field system of agriculture increased agricultural productivity and the diversity of crops grown. And it likely enabled the greater utilization of horses, which were better draft and transportation animals than oxen.[12] Over time there is no question that trade picked up steam in the shadow of greater agricultural surplus.[13] As trade expanded (both overland and by sea) the merchant class gained ascendancy. The merchant was as Robert Heilbronner put it "a disturbing yeast in the leaven of society."[14] This class was interested in making money by buying cheap and selling dear. Between the 1500s and the 1700s the expansion of trade picked up so much force that Jason Moore says one can discern a capitalist "world ecology" taking form. Colonial subjugation of native populations, slavery, and exploitation of the gifts of nature were part of this change.[15] In the same world systems tradition Lewis and Maslin likewise tell us: "In the sixteenth century everything began to change, and change with increasing speed . . . those living on the western edge of the continent of Europe changed the trajectory of the development of human society. . . . Nothing would be the same."[16] The conquest and pilfering of Africa and the Americas by Europeans proceeded with increasing speed. This process introduced into Europe sugar, slaves, cotton, potatoes, and lots of gold and silver among many other things, and it certainly expanded trade feeding the inherent benefits of buying cheap and selling dear.[17] To appreciate the enhanced interconnectedness that was formulating, think about the fact that "European silver, extracted from the Americas, was traded for Chinese luxury goods."[18]

Wealth was amassed by merchants, but that wealth would eventually fold back onto production; that is, it would change the forces and relations of production to become a fully formed capitalist system. First, feudal relations would have to be undermined and with it the manor as the focal point of production and distribution. And second, a new form of production and distribution would have to be established. As Eric Hobsbawm tells us: "The great frozen ice-cap of the world's traditional agrarian system and rural social relations . . . had at all costs to be melted" and in its stead a system of "profit producing private enterprise" installed.[19] The expansion of markets would move from buying cheap and selling dear to producing cheap and selling for less than your competition. Changes in the nature of production and not simply an expansion of trade and colonial subjugation would have to enter the mix.

One can follow this transition by looking at it through the lens of textile production, one of the most important industries in the early formation of capitalist production. Ground zero was Britain. Under feudal organization most textile production had taken place mainly on feudal estates and was a matter of processing, spinning, and weaving wool into cloth. The raw materials were produced on the feudal estate and so was the cloth, and its distribution was not done through the market but according to custom and tradition. But as international trade expanded so did the demand for textiles, and that demand folded back onto feudal relationships and the organization of textile production itself.

The landed gentry was pulled into the market system initially because of an increased demand for wool. Enclosure movements occurred over centuries in Europe but picked up steam in the eighteenth century. Common grazing land used by serfs and peasants was enclosed to raise sheep for the burgeoning textile industry. This was a long process of change that culminated in a "free" labor force with no other means of support other than to sell their ability to work for a wage. Those peasants and serfs with historical ties to the land found their material position undermined through enclosures. They were left adrift. Eric Hobsbawm tells us that in Britain "Some 5,000 'enclosures' under private and general Enclosure Acts broke up some six million acres of common fields and common lands from 1760 onward, transformed them into private holdings, and numerous less formal arrangements supplemented them." The Poor Law of 1834 then removed all protections that had previously existed for the increasing ranks of the poor and those dispossessed of ties to land.[20] Thus began the making of the working class, an absolutely essential element in the emergent economic system.

By the late eighteenth century in Britain there were an amalgam of landowners, rentier farmers, smallholders, and paupers who had been removed entirely from any ties to land.[21] The landed gentry either rented or sold land to peasant farmers who in turn hired those fully or partially dispossessed of common land to help with agricultural production, which could be sold in nascent grain markets to pay rent to the landlord. Exchange relations were gradually replacing feudal ties. Smallholders without sufficient land to be self-sufficient and those with no land at all were prime candidates for the ranks of the working class. Many of these fully disenfranchised ended up as farm laborers, but others became workers in the nascent textile industry. Some migrated to urban centers where they took up the role of unskilled laborers. The labor power of this class would be bought and sold as would the raw wool (and ultimately cotton), the thread that was spun, and the woven fabric. A different integrated system was emerging.

Textile production was fed by expanding international trading networks, which is apparent if we look at expansion of trade in cotton. The demand and production of textiles in cotton was a globalized matter. As Hobsbawm tells us: "Colonial trade had created the cotton industry, and continued to nourish it." Cotton could not be grown in Britain; it was imported from colonial outreaches and desirable for textile production (as opposed to wool or linen) for obvious reasons. Again in Hobsbawm's words: "The cotton industry was thus launched, like a glider, by the pull of colonial trade to which it was attached; a trade which promised . . . expansion, which encouraged the entrepreneur to adopt the revolutionary techniques required to meet it."[22] And cotton textile production was within the grasp of available technologies emergent at the time. Petty capitalists set up cotton mills on rivers where falling water could provide energy and by the turn of the nineteenth century "at least a thousand were scattered over several English counties."[23] In this way the transformation of economic society was beginning to take recognizable form. Ellen Meiksins Wood clarifies the parameters of this change: "Even later than the seventeenth century, most of the world, including Europe, was free of the market-driven imperatives. . . . A vast system of trade certainly existed, extending across the globe. But nowhere neither in the great trading centers of Europe nor in the vast commercial networks of the Islamic world or Asia, was economic activity and production in particular driven by the imperatives of competition and accumulation. The dominant principle of trade everywhere was 'buying cheap and selling dear.' There was no single and unified market, a market in which people made profit not by buying cheap and selling dear, not by carrying goods from one market to another, but by producing more cost-effectively in direct competition with others in the same market." Again Wood makes an important distinction: "The main vocation of the large merchant was circulation rather than production."[24] The imperative in the market system was to produce things so that the producer has a competitive edge, which means a move from buying cheap and selling dear to producing cheap and competing to sell at a lower price.[25] This important alteration was clearly apparent in the production of textiles.

Over time a new system of textile production was taking form.[26] Colonies provided the raw material (using mostly slave labor) and textile mills supplied the thread for weaving using wage labor. Slaves were bought and sold as commodities, unlike their working-class counterparts whose labor power was bought and sold.[27] Producing cotton thread occurred in textile mills powered by water and human labor. Weaving thread into cloth initially remained piece work done on hand looms in the homes of peasants who had retained some access to land. Weaving in the home was preferable to working in the cotton mills because here workers at least had some control over their work life. Weaving (unlike spinning) was still a skilled occupation. Yet the demand for their work depended entirely on market conditions. Weavers became part of the piece work system where thread and idiosyncratic loom parts could be provided to weavers by petty capitalists and finished cloth given back to them for sale in expanding international markets. The dehumanization inherent to the system was seen most clearly in slave markets but there existed varying degrees of servitude and even "free" labor was not really free when the choice was to work in the nascent factories or starve. The conditions of the working class during the eighteenth and nineteenth centuries were dismal and squalid at best. E. P. Thompson tells us that "manufacturers . . . in the mill communities could be seen to make riches in one generation," and it was clearly evident to the working class that their exploitation was making the mill owners rich.[28] And we should remember it was Engels's study of textile manufacturing in England that led him to write his powerful essay on the conditions of the working class there.

Textile production embodied all of the ingredients of nascent capitalism—expanding markets and ties to international trade, standardization and mechanization of production initially in spinning but eventually in weaving, entry of petty capitalists and their property rights, a natural placement along dispersed falling water that structured competition and provided energy, and a group of laborers to hire to work in the mills. There was plenty of competition as the money required to start a mill was not unreachable. Finally, there were robust international markets for both raw material and for finished textiles pushing things along. Buying cheap and selling dear had folded back on production, where the organization of production could provide a unique form of surplus: workers would produce greater value than they were paid, and capitalist by virtue of ownership would claim this as their own.[29] When they sold the product, this surplus would become profit; then profit would be fed back into production, enhancing the competitive edge of the capitalist. As Karl Marx tells us, the capitalist "shares with the miser the passion for wealth as wealth. But that which in the miser is a mere idiosyncrasy, is in the capitalist the effect of the social mechanism of which is he but one of the wheels . . . the development of capitalist production . . . compels him to keep constantly extending his capital, in order to preserve it."[30]

Over time the synergy of expanding trade, altered conditions of production, and sufficient energy in agricultural surplus, wind, water, and forests had melded into something distinct—a fully formed market economy. Fossil fuel did not figure into this initial mix of industrial production.[31] The petty capitalist was compelled by the system to compete, and of course this was mostly a race to the bottom for most human beings and for the more-than-human world. Competition and market expansion fed back on the organization of production. Factories were organized to get the most labor out of the workers. The fabric of material life began to depend on this complex reticulation. There was a profound interconnected whole that was taking form, a uniquely formed economic superorganism.

Humans had been disembedded from the rhythm and dynamic of the more-than-human world since the dawn of agriculture, but with agriculture as the mainstay of economic life they were at least tied together through the seasons and to a place. At least there was that earthly connection. In feudal society lords and serfs were tied together on land (that was not bought and sold for profit), and their relationship was not mediated by market forces. But to create a market system meant all connections of humans to each other and to place were secondary to the demands of the market and its imperative.

Material survival was no longer so clearly tied to cultivation but instead to the price of bread, the wage level, and whether products could be sold for more than the costs of production. There emerged a distinct reticulation of humans to each other around the market economy. Capitalist production was the source of wage labor for the worker, and the worker was the source of exploitation sufficient to render a profit for the capitalist, something the latter needed in the competitive battle to stay in the game. Neither could do without the other. Workers found themselves with only the possibility of wage labor standing between them and starvation. They were reduced to the lowest level to which they could be reduced and still guarantee that they would be able to provide a day's labor. Interdependence was expressed in the highly integrated work enabled through machine production where each person takes up residence as a cog in the machine of production, but it was also expressed in the capitalist/worker codependency. Even the peasants that rented land were bound not simply to grow crops but to grow crops to make a sufficient return so that they could pay their rent and have enough left for themselves. This required hiring farm laborers and paying them as little as they could. All of this was in play by the time the use of fossil fuel entered the picture.

Karl Polanyi best articulates what was happening when he tells us, "To separate labor from other activities of life and to subject it to the laws of the market was to annihilate all organic forms of existence." Those organic forms, of course, consisted of being attached to community, family, and place. He draws out a similar distinction in the human relationship to land: "The economic function is but one of many vital functions of land. It invests man's life with stability; it is the site of his habitation; it is a condition of his physical safety; it is the landscape and the seasons. We might as well imagine his being born without hands and feet as carrying on his life without land."[32] Let's be clear: the dawn of agriculture began the process of separating humans from an existence embedded in the rhythm and dynamic of the more-than-human world where they were one of many, and it altered the relationship of humans to each other and Earth. Yet what transpired with capitalism was clearly a further distancing of that distancing, a more profound duality. No longer would a person even have a reliable place to reside, nor a connection to community or perhaps family, or even a connection to the rhythm of the day. And land became further disarticulated from its own organic whole. Just as humans were disembedded from community so too was land disembedded from its "ecological" community as the demand for raw materials began to expand. Ecologies of Earth where humans were but one of many were first replaced with agriculture's ecology where the economic superorganism took hold and then with the particular ecology of capitalism.[33]

It is difficult to say what might have happened had nascent capitalism not been augmented with a form of energy that appeared to be unlimited. It was, after all, an expansionary system. It would certainly have remained more modest and contained. An economic system, however institutional, is always undergirded with a biophysical foundation, energy being central. There is no question that access to a particular form of energy, first in coal and later in oil and natural gas, came to define the gears of exploitation and interdependence and how the expansionary dynamic of the system would play out. Ironically, fossil capital would allow the pulse of the economic superorganism to become ever more removed from its connection to Earth because access to energy appeared to be unlimited. Capitalism would eventually become the system of mass production and corporate entities and one where the belief in human ingenuity and markets to solve problems would become ideological bedrock. There would be no sense of absolute limits to the process of profit and capital accumulation.

Fossil Capital

Oddly, it wasn't a lack of energy that drove textile production to fossil fuel because early textile production was well provided for by the power of falling water, which was a reliable and abundant source of energy. The problem for early capitalism in textile manufacturing apparently came from lack of an accommodating labor force, an essential ingredient in production. Andreas Malm tells us, "For the mills growing along the riverbanks in the late eighteenth century, wage labour on a voluntary basis was not a sufficient option."[34] Spinning required workers for factories that were located where falling water was available, but unfortunately these were also places where it was difficult to access sufficient labor. Potential workers who had managed to retain a small plot of land—and thus could weave in their homes—preferred this option to spinning in the textile mills. Many of the factories located in remote areas turned to child labor imported from the poor houses, a strategy that became increasingly abhorrent. According to Malm, steam was eventually utilized as a preferred option to falling water not because it provided superior energy but because it offered "a superior method of energizing the exploitation of labour."[35] In this fact we can garner greater insight into the inner workings of the emerging system.

Production could be transported anywhere with the use of steam power fueled with coal, and it was better to produce in urban settings with more "free" laborers depending solely on wages. Immigrants (primarily Irish) and those who had lost access to land congregated there, as did many skilled artisans and workers in the building trades. Urban enclaves also offered opportunities for unskilled labor—loading and unloading ships and transporting coal for heating and cooking among other things. According to Lynn White Jr., "From the Neolithic Age until about two centuries ago, agriculture was fundamental to most other human concerns. Before the late 1700s there was probably no settled community in which at least nine-tenths of the population were not directly engaged in tillage."[36] Yet by 1840, 40 percent of the working class in Britain was engaged in industrial production fueled not by falling water but coal, and industrial production came to be located in urban settings. Malm tells us that "the first half of the 1820s marked the record of 2.6 percent annual increase in the urban population of England . . . Manchester swelling with an average of 3.9 percent in the 1820s."[37] And certainly Britain urban centers had the added advantage of easy access to global markets since virtually all were located with access to water transport.

The use of fossil fuel in textile production ultimately enabled the joining of spinning and weaving. Given the bottlenecks between spinning and the piece work of weaving, technologically integrating them became important to ensure a continuous and reliable supply of textiles as markets expanded. Again I refer to Malm's comments: "As long as traditional weaving flanked mechanized spinning, the insertion of a mill in a rural backwater remained feasible or even expedient, as surrounding households were often engaged in the industry through putting-out. A combined factory, on the other hand, increased the need for operatives *and* cut the ties to the web of semi-agri-cultural, semi-domestic workers that populated the northern country side, at once raising the value of spatial concentrations of people and reducing the need for outworkers. Like the self-acting mule, the power loom demanded of its minders utter resignation to the diktat of the machine. It pulled the industry to the town with redoubled force."[38] Fossil capital carried textile production to another level of rationalization and integration where the costs of production were cheapened by accessing a larger pool of laborers but also by deskilling and pacing the workers thereby taking the last vestiges of control out of the hands of the workers (in this case the weavers).[39]

Of course, it wasn't simply textile production that was the target of industrialization and rationalization using fossil fuels. Any industry was the target but in all cases the results were the same. There is a logic in the way capitalism organizes production and this logic becomes more extensive with the use of fossil fuel. The process of utilizing large-scale mechanization and orchestrating a more detailed and integrated division of labor is a defining feature of fossil capital. E. P. Thompson comments on the result of the system in its early years: "The first half of the 19th century must be seen as a period of chronic under-employment, in which the skilled trades are like islands threatened on every side by technological innovation and by the inrush of unskilled or juvenile laborer." This, of course, was not universal as technological change "devalued old skills and elevated new ones," but as a general rule deskilling was a trend that accelerated with the use of fossil technology.[40] Deskilling through technology was a standard part of the process of exploitation that itself was necessary to the goal of profit making under competitive market conditions. The last thing a capitalist wanted (wants) was (is) workers with any power over production, and they certainly didn't want workers who demanded higher wages and had the power to get them based on strategic knowledge and skill. This imperative only increased with the greater investment in machinery. The rationalization of the labor process was extended with the use of fossil fuel in accordance with the imperatives of the system—making profit in the context of competition with other capitalists doing the same.

Fossil capital extended the productive potential and expansionary dynamic of the system, which became explosive. The exponential growth curves that began their marked propulsion into the stratosphere at the beginning of the nineteenth century are testament to this fact.[41] In fact, so dramatic was the effect of fossil fuel on the expansionary dynamic of capitalism that most people forget that the foundational fabric of capitalism, in its particular form of surplus in profit and the centrality of markets in economic life, was first woven without the benefit of fossil fuel. David Landes captures the explosive nature of the system with the use of fossil fuel when he writes, "By 1800 the United Kingdom was using perhaps 11 million tons of coal a year; by 1830, the amount had doubled; fifteen years later it had doubled again; and by 1879 it was crossing the 100 million-ton mark. This last was equivalent to 850 million calories of energy, enough to feed a population of 850 million adult males for a year (actual population was then about 31 million)."[42] It was nothing short of remarkable.

Energy is necessary regardless of the particular tapestry of material life, but some systems expand the energy imperative and others do not.[43] An agricultural system, unlike hunting and gathering, is an expansive and energy-seeking system. The expansive nature of the market system with its surplus in the form of profit would have eventually run up against energy limitations had it not been for fossil fuel because the feedback loops between profit, production, and expanding markets create a self-reinforcing expansive spiral and this requires energy. Forests, wind, water, and agricultural output would not have accommodated the inherent dynamic of capital accumulation in the way that highly dense carbon did. Ironically, fossil fuel freed market capitalism from any energy limitation unleashing the full force of the expansionary dynamic inherent to a system where surplus has taken up residence in exploitation and garnering of profit.

The expansionary dynamic of a capitalist system cannot be reduced to a simple energy matrix, but there is no question energy is accommodating to this expansionary interdependent system. An abundance of fossil fuel will not expand capitalism unless it is profitable to produce and profitability under capitalism was (and is) determined by the ability to sell and to sell for more than it cost to produce within the particular competitive structure that exists. Depressions occur when there is plenty of energy, and depressions are reliable in this system. The probability for overproduction (or underconsumption) is clearly great in such a system and so is the need to continuously expand markets. To see more clearly the tensions in the system and how they are exacerbated with fossil capital think of it in this simplified narrative. There is a logic in the system to deskill workers as wages can be minimized and greater control over the process of production both increase the competitive edge of the individual capitalist. But to cheapen labor means it will be hard to sell all that is produced; and there is a lot being produced—as there are abounding economies of scale in fossil capital. Massive quantities of low-cost goods are produced. A tension lies here—the capitalist wants to pay the worker less while needing them to buy more. So while the impulse of the system is to exploit and cheapen labor, the system relies on exploited and cheapened labor to buy the products that are sold in order to realize profit. In simplistic terms the pursuit of profit, especially after it is augmented with fossil fuel, establishes its own contradiction.[44] Should it surprise anyone that in its more advanced form advertising and the extension of credit are both necessary to the system as is the continual expansion of markets.

This is but one version of the complexity of the system that helps to bring our understanding of the system into sharper focus. There are others. In this rendition it isn't simply tremendous growth that defines the system, it is growth defined by the continual threat of stagnation, and growth and stagnation are matters internal to the system. It is also a system where tremendous economic power resides with the few.[45] Yet workers and capitalists remain interdependent, and profit should never be reduced to a simple dictum of greed. Clearly things are much more complicated than that. The system logically moves from a simple "collection of small powerless companies who accept passively the impersonal force of the market and forego large economic profit in the interests of low consumer prices and stable general equilibrium" to a much greater concentration of economic power where domination "by giant corporations . . . that would seek to control the forces that undermined this position."[46] Indeed, in the shadow of stagnation those capitalists that survive the continuous weeding-out process will become more powerful, but their victories will never completely secure their position because the conditions of their supremacy are constantly in flux.

Given the complex landscape of this system, it shouldn't surprise us that economic discourse has been preoccupied with sorting its "interior" life. The energy source that now generates the specter of climate change accommodated the emerging system and brought the logic and tensions of that system to full fruition. Fossil fuel freed the system to become fully itself. At the same time market capitalism is a material system and the demands of the system on the Earth are continuous and cumulative. This is a paradox of economic life that we have difficulty understanding (and navigating). Capitalism is at once a supra-material system and yet profoundly material. This is the extreme form of duality between humans and Earth that has taken hold. The duality cultivated in the ten thousand years since the advent of agriculture has reached its apogee with fossil capital. The system's interior life eclipses its biophysical foundations even as the latter is crumbling. The problem for the system and its relationship to Earth is not simply that capitalism uses nature in its continual dynamic of capital accumulation and expansion, but again, that it is at once both a supra-material and a material system. This is the full expression of the duality that began with grain agriculture and the formation of the agricultural system.

The denial of the implications of this profound duality is continually reinforced by the human capacity for inventiveness and the belief that the market channels this propensity to solve problems of temporary bottlenecks. Limits are never perceived as absolute; they are always just momentary processes of readjustment. With the specter of climate change before us, redemption is imagined in the form of an expanded green energy system, a new technology channeled through the market that will provide jobs, and a new sustainable form of energy for an ever-expanding system. Of course, this is an elaboration of duality; it does not counter it. The belief in technology and its endless efficiency gains, in progress (mostly encouraged through the market), and in our ability to manage the egregious deficiencies of the system without fundamentally altering the system are the ways we retreat from the paradox of our historical moment and the challenge of real rapprochement with the Earth.

Chapter Seven

In Search of a Deep Ecology of Economic Order

Capitalism is a system that functions as a supra-material system as well as a material system. An economic system is always a material system—it has a material relationship with the Earth: coal is mined and burned, iron ore is extracted and smelted, crops are grown in soil, animals are raised and eaten, forests are cut, wastes are emitted, and so on. But capitalism also functions as a supra-material system where economic variables interact as if disconnected from the Earth. An example will help make this clear. Think of some of the common economic variables used by economists in standard macroeconomics: wages, profit, rent, income, spending, investment, savings, inflation, and GDP, for example. In a typical macroeconomic model taken from any introductory textbook the economy is presented as a circular flow of income being generated in the production of goods and services and then being spent on those goods and services. If this relationship is interrupted; that is, if not all of the income is spent, there is a contraction of the economic system. There is nothing in this model of the economic system that refers directly to the Earth. More importantly, it is not limits imposed by the Earth that causes the contraction but rather a disconnect between income and spending. The Earth is assumed to give forth when the economic incentives are there, and if there are any problems with limitations of the Earth, human ingenuity and price signals are assumed to solve those problems; substitutes will be found and new technologies developed. And if a bit of help is needed—government policy can tweak the system.

Over the past 250 years the ideas and preoccupations of the "worldly philosophers" have orbited this supra-material realm and relegated the connection of the economy to Earth to the margins of their analyses. I borrow the term "worldly philosophers" from the book of that name written by Robert Heilbronner in which he discussed the lives and ideas of the great economic thinkers.[1] In this sense, a social science constructed to understand the material organization of human society mostly lost sight of the centrality of the earthly roots of the economy. The worldly philosophers might more accurately be called *unearthly philosophers.*

I am not one to defend the unearthly philosophers for their oversight, but it is essential to understand that capitalism is a complex and dynamic system that has matured as a supra-material system. The self-referential inclination of human economic life since the rise of grain agriculture came full circle when capitalism was fertilized with the industrial revolution and the use of fossil fuel. The profit system appeared to have no earthly limits, and an earthly disconnect framed the foci of the analyses of the unearthly philosophers. And among those who have tried to understand the complexity of capitalism as a supra-material system, there is no consensus about it. It is a system of internal contradictions, instability, and conflict to some, and one devoid of contradictions and conflict—a well-oiled machine to others. It is the constellation of these differences that has occupied the narratives and debates of the economists. While the failings of neoclassical economics (the dominant economic paradigm) have been commonly noted among that small chorus of economic voices now trying to connect economy to Earth, the truth is that most economic paradigms put forth over the past 250 years fell victim to the same deficiency—the connection of economy to Earth has not been central to their analyses. We do understand more about capitalism as a result of the debates of the great economists. And in a very real sense, their intellectual dispositions merely reflected a real duality between economy and Earth. Their focus was the supra-material life of capitalism. A brief review of the ideas of the unearthly philosophers will help to clarify my proposition.

The Unearthly Philosophers

This is not intended to be an expansive journey into the ideas of the great economic thinkers and how they handled (or did not handle) the connection of the economy to Earth. I only want to show that the connection of the economy to Earth orbited the outer reaches of the galaxy of their thought.[2] The debates of the unearthly philosophers revolved around growth, income and its distribution, trade, prices, the accumulation of capital, stability, instability, exploitation, equilibrium, efficiency, depression, debt, interest rates, and inflation, for example. For illustrative purposes I briefly chart the foci of Adam Smith, David Ricardo and Thomas Robert Malthus, Karl Marx and Friedrich Engels, the neoclassical economists (Léon Walras, Carl Menger, and William Stanley Jevons), Thorstein Veblen, and John Maynard Keynes. These are considered to be some of the foundational economic thinkers in the history of economic ideas, and they are the same economic thinkers that Robert Heilbronner highlighted in his book.[3]

Begin with Adam Smith. Adam Smith's *Wealth of Nations* didn't "fire the gun" of our present race, but it did frame one understanding of the system at play.[4] Writing at the end of the eighteenth century, Smith clearly understood that beneath the chaos unfolding around him (and it did appear chaotic) there was an order and system integrity that he attempted to decipher. Smith held that market expansion was necessary to lift humanity out of an impoverished existence. He was not concerned with the question of limits except to note that limits were a product of the extent of the market. He was preoccupied with the idea that capitalism could expand the productive potential of human society. In his mind it was a rising tide that would lift all ships.

Like many economists, Smith began with certain assumptions about human nature. Among the traits inherent in all humans were the propensity for self-love and a natural propensity to truck, barter, and exchange. Smith considered the propensity to truck, barter, and exchange a form of self-love in the economic realm as humans always and everywhere traded "this for that" in order to better their own position. Smith believed the propensity to truck, barter, and exchange would play off of the division of labor (they were like a hand and glove), and the division of labor would make humans more productive. Smith's words say it best: "The greatest improvement in the productive powers of labour, and the greatest part of the skill, dexterity and judgement with which it is anywhere directed, or applied, seem to have been the effects of the division of labor" and "This division of labor, from which so many advantages are derived, is not originally the effect of any human wisdom. . . . It is the necessary . . . consequences of a certain propensity in human nature . . . the propensity to truck, barter, and exchange one thing for another."[5] In Smith's thinking the productive potential of labor elaborated through a division of labor and encouraged by the natural propensity to truck, barter, and exchange had taken exaggerated form in capitalism where the competitive market (the invisible hand) organized and disciplined economic life. In this way Smith thought capitalism was the natural order of society. The invisible hand would order self-interest and promote an ever-greater division of labor: social welfare and economic expansion would be the result. All that was needed was to allow this system of "perfect liberty" to take hold and expand. The only thing that would stop the system was a limit to the expansion of markets. Adam Smith stood barely at the cusp of the industrial revolution, which was just taking hold as the ink was drying on his tome. World population stood at under one billion people. I often wonder what Adam Smith might have written if he was standing here now.

Next consider David Ricardo and Thomas Robert Malthus, who were contemporaries writing in the late eighteenth and early nineteenth centuries. Conversations about distribution of income and its effect on economic growth (accumulation) dominated the discussions between them. At the beginning of the nineteenth century, income was distributed to three different classes. Rent went to the landed gentry, profit to the capitalist, and wages to the workers. There was concern about the implications of this distribution for the expansion of the economic system. The important discussion between Malthus and Ricardo centered on a debate about whether the Corn Laws should be eliminated. Ricardo believed that if trade in grains was limited with the rest of Europe, Britain would be forced to expand agricultural production onto less fertile land causing both rent and the price of food to increase. Wages would increase, and profit would fall. This would reverberate through the system and effect the expansion of the system because profit was the source of investment spending, which fueled the expansion of the economy according to Ricardo.[6] Malthus was not as concerned as Ricardo with the problematic effects of any redistribution of income away from profit and toward rent. Malthus recognized the possibility of "gluts" and believed that the consumption spending of the landed gentry, though not considered productive, would bolster the system even if it was spent superfluously (that is, not in investment in more factories and businesses but rather more servants). Ricardo's remedy for the diminishing returns to agriculture and its effect on the economy was to expand trade in agriculture thereby limiting the effect of the diminishing returns in agriculture. And as earthly as Malthus might have seemed in his observation about population outstripping food supply, his major concern was not with the issue of absolute limits but instead following the internal logic of the system and the effect of different policies on the distribution of income and how that might work its way through the system.[7]

Karl Marx (and Friedrich Engels) engaged an analysis of capitalism using the methodology of historical and dialectical materialism. It is the organization of humans in the reproduction of their material existence that determines history, and this is especially complex for humans because they produce their material existence socially, and they are intelligent. While Marx had a great interest in history his understanding of human prehistory was limited. His focus was clearly on human societies producing surplus (class societies) and more narrowly on capitalism, its origins, ongoing unfolding, and the contradictions and tensions that arose therein. These contradictions and tensions were not primarily couched in relation to Earth but were contradictions and limits internal to the system, which appeared as a tension between the working class and the capitalist class. A brief reiteration of his thesis (not an exegesis) goes like this. Human labor produced all value. The pursuit of profit depended on exploitation of workers at the point of production. Exploitation (the extraction of surplus value or not paying workers the full value of what they produced) was the source of profit, and profit was necessary for the accumulation of capital. If capitalists wanted to stay in the game they were constantly compelled to exploit, profit, and reinvest. Yet that process was fraught with contradiction. Workers could be ever more efficiently exploited through technological change (think mass production assembly line work), but the increased productivity of the system, and greater inequality, led to a greater probability of under-consumption or overproduction, thereby limiting profits in the future. Thus the system would be plagued by class conflict and overproduction, which led to depression. It was not the prospects of reaching ecological collapse or biophysical limits that were foremost on the mind of Marx, it was the internal contradictions of the system.[8] In this way capitalism would sow the seeds of its own destruction.

The neoclassical economists (Léon Walras, Carl Menger, and William Stanley Jevons) were clearly more concerned with presenting a system of harmony and order and giving it mathematical integrity and the mark of science than framing their theory around the problems of the day.[9] In this they parted ways with the classical economists (Smith, Malthus, Ricardo, Marx and Engels). They published their ideas (all similar) in the last quarter of the nineteenth century at a time where capitalist production was becoming increasingly concentrated and the rise of the corporation was at hand, and yet the core of their refined system analysis allowed for little of that reality to creep in. The neoclassical economists perfected a systems analysis in the spirit of the legacy of Adam Smith (competitive markets and self-love) and J. B. Say who told us that supply creates its own demand. In other words, the production of goods creates income that is spent on that production, and any imbalance between supply and demand will be corrected through price movements so that markets will always reach equilibrium. Theoretically gluts and unemployment were only temporary. The neoclassical economists added Newtonian physics (a movement from equilibrium to equilibrium), and the modeling of relationships through equations and axiomatic assumptions thereby giving economics the stamp of science. The invisible hand would orchestrate the movement from one equilibrium to another while maximizing social welfare all the while promoting individual liberty.

All that was necessary to achieve the results of neoclassical theory were a few simplifying assumptions about human nature (that humans are self-interested and driven by the pursuit of pleasure), market structure (markets are perfectly competitive, meaning no individual firm or consumer has any power in the market), and a total disregard for the movement of economy over the course of history. In the neoclassical framework human economic behavior (rational economic man) was universal and unchanging over time, rendering culture and history irrelevant. And no matter that the economy had moved far beyond the butcher and brewer on every street corner—the economic landscape of Adam Smith's world. No matter that fossil capital and the maturation of the system had changed the boundaries of competition and the dynamic of accumulation and capitalism was entering an oligopolistic stage. No matter that workers and capitalists were in conflict. In the neoclassical world, harmony and efficiency abounded. Workers and capitalists would get a reward in the form of wages and profit commensurate with their contribution to production, and social welfare would be maximized if the market was left to its inclination. The only limits to the system would be temporary, and these would show up in price movements that would provide the signal that induced change and adjustment. The market would capture the benefit of human inventiveness and ingenuity because there were economic gains to be made.

Neoclassical economics gained ascendancy because it had ideological appeal, and it appeared to shine a scientific light on the system of perfect liberty sanctioning it in a more forceful way. Once the assumptions are accepted, its logic is impeccable. As with other paradigms the entry of the Earth into this analysis exists only around the margins: an afterthought. Shortages would show up in prices, and adjustments will be made; thus there is no threat of absolute limits, only temporary shortages. Eventually some neoclassical economists recognized the presence of "market failure" such as externalities highlighted by A. C. Pigou. But externalities and other sorts of market failures were (and are) considered aberrations in an other-wise well-functioning system in a neoclassical framework. These aberrations could presumably be dealt with through policy—for example a Pigouvian tax (e.g., a tax on carbon).

Thorstein Veblen, an American economist writing in the late nineteenth and early twentieth centuries, understood that the economic system was not a movement from one equilibrium to another but instead unfolded over time. Like Marx and Engels, Veblen saw the economic system in historical context. He attempted to capture change but also to connect the economic system to universal human tendencies. Veblen grounded his analysis of capitalism in an interpretation of human nature that ran in contrast to "economic man" and his assumed natural propensity to truck, barter, and exchange. Veblen believed humans exhibited universal behavioral tendencies or clusters that he labeled the "instinct of workmanship" and "the predatory instinct." The instinct of workmanship encapsulated inventiveness, the ability to work together, the inclination to nurture: while the predatory instinct was the tendency to exploit. Racism, sexism, and ultimately the capitalist profiting were all manifestations of the predatory instinct. The form these inclinations took varied over time, but they were universal tendencies.[10]

Veblen's most important contribution to the analysis of capitalism came in the form of a belief in the tension between human ingenuity, the ability to work together, and the power of technology to make human life better on the one hand (all embodied in the instinct of workmanship) and the pursuit of profit on the other (the embodiment of predatory instinct). He saw these at odds and especially apparent in the ruinous competition among large corporations taking place at the end of the nineteenth century. That competition created the inability of society to take full advantage of its technological and cooperative possibilities; that is, to provide humans with cheap products efficiently produced. To maximize profits, corporations would withhold efficiency in order to bolster prices, and they had the power to do so because contrary to the neoclassical world the economy was not made up of powerless firms.[11] The result was unemployment and unused productive capacity—in other words, continual depression.

Veblen's penchant for understanding the cultural manifestations of his clusters of behavior (which he thought were expressed uniquely during different historical periods) led him to his disquisition on consumer culture. But he wasn't interested in consumer culture in the sense that consumerism would outstrip the ability of the Earth to support it (although later scholars of Veblen have used it in this way). Consumer culture was simply a way to broaden economic analysis and to understand that there are cultural elaborations of an economic system that reduced the tensions in that system. In his analysis, the class tensions in the system were moderated through consumer culture where individuals were preoccupied with keeping up with the Joneses, rather than trying to figure out why the Joneses had so much. Despite his Wisconsin farming background Veblen stood squarely in the camp of unearthly philosophers.[12]

John Maynard Keynes also saw a system of internal tensions that would result in crisis. But he did not see the inevitability of revolution proposed by Marx and Engels. Instead Keynes saw an economic system that could be managed and the damages of its worst tendencies contained. Keynes believed that supply might not create its own demand, and markets might not clear (prices might not move to take care of gluts and unemployment), so prices couldn't be relied on to bring markets back into equilibrium once they got off kilter (and they reliably would in an advanced capitalist system). Depressions were inevitable.[13] Keynes's basic idea about depression was this: there were leakages out of the circular flow of income and spending that were not offset by injections back into the spending stream. The problem of leakages being greater than injections caused the system to contract, and a contraction would build on itself in a multiplying fashion. It wasn't a biophysical problem that would create crisis; it was, for example, the leakage of savings not being offset by the injection of investment spending. In other words, there would be insufficient aggregate demand. Keynes considered investment spending to be particularly problematic because it was based on expectations (animal spirits) that ran especially bad in a time of economic contraction. It would take the government intervening to right the economic ship because the private sector would not invest in bad economic times, and consumers who had lost jobs had no money. The government could utilize fiscal and monetary policy to stimulate the economy. Job creation would invigorate spending, which would eventually reinvigorate investment.

The legacy of Keynesian economics has been institutionalized over the past century. The expansion of social welfare institutions has helped to moderate the chronic problems of inequality and poverty, unemployment, and the instability of the system. Social Security, the Full Employment Act of 1946, unemployment compensation, Medicare and Medicaid, minimum wage laws, and various forms of welfare payments are all examples of institutional arrangements used to stabilize the system. Amid the destabilizing effect of the

COVID-19 pandemic we have seen panic emerging around the economic contraction. Without spending and jobs the economic engine will come to a screeching halt, and this will multiply on itself. Government stimulus is a necessary part of the economic landscape of the moment, and Keynesian policy and institutions are as foundational to twenty-first-century capitalism as they were to twentieth-century capitalism. Despite the swerve to the right with the rise of neoliberalism, we have an economy that is still bolstered with these methods of managing the instability of capitalism. But let there be no confusion about this—Keynesian policy to deal with stagnation and inequality is expansionary. It brings into sharp focus the dialectical tension of the economic moment. Inequality, poverty, and instability of the system are dealt with through expansionary Keynesian policy. And the problems of the relationship of the economy to the Earth need the opposite.

Over time we also have developed particular institutional arrangements to manage environmental problems and resource use with the assumption that market failures occur, and the government has a role to play when markets fail. Policy takes form in the institutions that oversee the economy and environment interface—the EPA, the Department of Agriculture, the Department of the Interior, and the National Resources Conservation Service are but a few examples. All of these are agencies that attempt to manage the problematic outcomes of capitalism and Earth while holding to the belief that the relationship can be managed while adhering to the present system. These are not revolutionary; they only appear so in the shadow of the assault on the Earth brought forth by neoliberal economics.[14]

It is essential to keep in mind that the rise of neoliberal economics is a problematic sideshow to a more fundamental tension and duality that both liberal and conservative economic thinkers fail to acknowledge. There is no question neoliberal economics has enabled the war between the economy and Earth to become ever more apparent, but it didn't create the war. Even without neoliberalism the challenge of the dialectical tension between the need to "solve" problems of inequality, poverty, and stagnation and the need to ensure ecological health of the planet has been ineffectively navigated by policy and the economic thought that undergirds it. We are a long way from resolving the problems of inequality and overshoot simultaneously, and the dominance of the unearthly economists continues to hold sway even as we cross planetary boundaries and languish in the paradox of duality.

The confluence of the challenge of inequality and ecological decay demands a foundational shift in economic thinking: one that places the connection between economy and Earth as the pivotal axis of analysis.

This isn't the first time the earthly reality of economic order has raised its earthly head. Indeed, the ecological effects of the economic superorganism on the Earth are as old as civilization itself.[15] Yet the collective impact of an expanding economy with an inherent tendency to profound inequality and instability, the sixth mass extinction, the reality of climate change, and a global population of almost eight billion people create a particularly problematic moment in the trajectory of the economic superorganism. The damages to the Earth have taken flight on the upward neck of exponential growth curves. At the same time, the demands of the impoverished masses accelerate. The gap between the have and the have nots becomes ever more problematic to navigate with old recipes of growth. Economic thinking must enter the realm of the historical moment and confront the apogee of the economic superorganism. If it is to be relevant it must provide a lens for adequately processing our predicament—a lens that gives a clear understanding of where we stand and what rapprochement between economy and Earth means.

Whether ideas have the power to alter the trajectory beginning with the agricultural revolution is questionable. That is not a question taken up here. But economic discourse should at least help us to interpret our circumstances in such a way that the right questions are asked of the historical moment. For example, the climate challenge is not simply a question of how to deploy ever greater quantities of renewable energy but rather how to engage the challenge of limits. The challenge of sustainable agriculture is not how to feed ten billion people but instead how to produce food by listening to "the genius of place" and conserving our soil.[16] And the challenge of maintaining ecological integrity and creating a sustainable economy is not a question of how we manage ecosystems but instead how we take our place as but one of the many species that inhabit Earth.

The Rise of the Earthly Philosophers

The historical moment demands the rise of *earthly philosophers* whose ideas can help us ask the right questions and correctly frame the present challenge. This is a quest that must go beyond the confines of crude sustainability and to the heart of human existence and the relationship of humans to Earth. Nothing less rises to the challenge of the historical moment and the sixth mass extinction. An economic system frames and contextualizes human material life and the relationship of humans to the more-than-human world. We must now come to terms with the apogee of a fundamental change in economic life that began more than ten thousand years ago. A day of reckoning has arrived as we stand at a juncture in the history of *Homo sapiens* where humans are taking a final and irrevocable step along a trajectory that began with the agricultural revolution. Human beings are becoming irrevocably chiseled into *Homo sapiens agriculturii*, a wholly domesticated species on a wholly domesticated and humanized and degraded planet. In this sense the ecological crisis is also a crisis in the history of the evolution of humans.

Interpretation and focus are crucial—the lens of economic thinking must move us in the direction of the quest for a deep ecology of economic order. The unearthly philosophers have been captured by the supra-material economic system, but there is a new ilk of economic thinking taking form. Let's label those who occupy its ranks the *earthly philosophers*. Do their ideas lead us to a rapprochement between economy and Earth sufficient to the historical moment—to stave off the sixth mass extinction for example? Do they sufficiently register duality and the turbulent waters of the paradox of the historical moment where we need growth for jobs and degrowth for the ecological health of the planet? Do the earthly philosophers fully understand that humans stand on a divide in their evolutionary history?

I divide the emerging voices of the earthly philosophers into two groups although in some sense this demarcation is arbitrary. All, in one way or another, focus on the problematic relationship of humans to Earth that emanates from our economic system. One can argue that there is enough integrity in each group to categorize it as distinct, although there is certainly crossover and intermingling. I label these ecological economics (EE) and ecological political economy (EPE). In each there is by now an extensive body of literature, and they clearly stand outside the main current of economic thinking, stuck as the latter is in the language and orientation of a supra-material economy. Again, the intent here is not a survey nor an exegesis but an introduction to some of the main threads of their thought.

Ecological Economics (EE)

In its attempt to connect the economic process to the Earth, ecological economics (EE) begins with a simple materialism.[17] It presents the economy as a subset of a larger Earth system and points out that infinite growth on a finite planet is impossible. The many connections to the Earth are then highlighted and elaborated. This is descriptive and accurate; it is impossible to have infinite growth of an economic system that is a material system on a finite planet. In this EE offers a simple and intuitive proposition that is easily visualized and resonates with those concerned about the Earth: unlimited growth on a limited Earth is not possible. Yet beneath this simple logic there is a lot of complexity that EE only begins to unravel. The economic system cannot grow indefinitely—this is simple and so is pointing out that the economy is a subset of the Earth. But asking how and when growth ends is infinitely more complicated as is fully registering the problematic duality and paradox of the system (that it is simultaneously supra-material and material and that it requires both degrowth and growth).

My own sense is that a foundational exploration of the complexity of our circumstances among ecological economists has been slow to emerge because they settled into a simple materialism rather than a more expansive materialism. A more expansive materialism would have enhanced their insight into the complexity of the human relationship to Earth and its contextualization in economic systems.[18] EE has ended up descriptive (the economic system is a subset of the earth), its solutions incomplete (put a price on natural capital or a cap on resource use without the sense for the need to change the underlying relations and conditions of production). And a sense for the importance of a deep ecology of economic order, one that seeks to fully understand the human/Earth relationship that comes to rest in economic systems, has remained underdeveloped.

In order to illustrate the approach of EE, I take the example of two of its seminal thinkers: Nicholas Goergescu-Roegen and Herman Daly. Again, my purpose is not an expansive survey of their work nor of EE. I simply want to illustrate the way the focus of these two scholars set EE on a certain path. I begin with Nicholas Georgescu-Roegen, who coined the term "bioeconomics" to demarcate a new approach to economic analysis. In this he wanted to highlight "the biological origin of the economic process and thus spotlight the problem of mankind's existence with a limited store of accessible resources, unevenly located and unequally appropriated." According to G-R the human species has an underlying imperative for energy and materials (as do all species), and G-R introduces the Entropy Law in order to help think about the material connection of the economy and the limits faced. He also tells us: "The Entropy Law is the only natural law that does not predict quantitatively. It does not specify how great the increase should be at a future moment or what particular entropic pattern will result. Because of this fact there is an entropic indeterminateness in the real world which allows . . . life to acquire an endless spectrum of forms."[19] He is right about that, but he did not wade into the question of entropic forms and more specifically how different they might be for humans over time.

The strength of G-R's contribution was not in his exploration of "entropic indeterminateness" but rather in demonstrating the connection of life to its use of materials and energy and more importantly the irreversibility of the use of energy dictated by the Entropy Law. Thus, his work highlighted the way the human economy was connected to Earth in a very basic material sense that especially resonated in the era of fossil fuel use and emerging awareness of the limits to growth. The human species is sustained at the most fundamental level through available energy and matter because "the economic process is not an isolated self-sustaining process." In this way G-R moves economic thinking away from what he called its *timeless kinematics*."[20] This is a simple but irrefutable materialism that the unearthly philosophers with their focus on the supra-material had mostly ignored.

Georgescu-Roegen understood that humans are somewhat unique in their ability to change their relationship with Earth through their ability to make tools or what he calls "exosomatic instruments." (I have no disagreement except to note that some animals do the same.) He argues that in the process of toolmaking, humans developed a big brain, which furthered their inclination toward inventiveness (exosomatic extensions) and what he calls an "addiction to their inventions."[21] If by this he means there is a dialectical interplay between humans and Earth and out of that humans became a smart, inventive species he is not wrong (many scholars have said the same). Georgescu-Roegen interprets inventiveness (our addiction to gadgets) as a positive feedback loop that leads to growth and social complexity. Georgescu-Roegen has stated that "the roots of economic growth lie deep in human nature," and he adds that hierarchy is often part of that complexity where "social mythology has always been erected . . . to justify the abusive growth of special privileges."[22] This is G-R's explanation of why humans became the expansionary and class-based species they are. There is no particular theoretical investment in acknowledging major entropic revolutions or explaining why, for example, humans were embroiled in a stable, no-growth economic system for most of their history. Rather there is a continuum of human inventiveness, growth, and social complexity.

The analysis started by G-R went in the direction of a focus on energy and the Entropy Law as a way to connect economic systems to Earth and to understand the limits of the economic system set therein. It did not go in the direction of explaining more completely the formation and dynamic of particular entropic systems nor how they might differ qualitatively and in relation to Earth. Again, his idea that there is a constant struggle for energy fits with contemporary preoccupations with energy. While it is true that all life needs energy, the long arc of human history tells us that *Homo sapiens* lived for two hundred thousand to three hundred thousand years in modest numbers, utilizing very modest technology and at the same time adapting to dramatic climatic changes. By all accounts theirs was not a relentless struggle for more energy although there is no question they needed it. In other words, it appears *Homo sapiens* were engaged in some measure of entropic equilibrium over a very long span of time.

Georgescu-Roegen's inclination that economists should "turn away from ill-fitting positivism of the past hundred years and . . . start looking at the economic process from a physiological and evolutionary viewpoint in a dialectical manner" has never reached full promise, but he did help to sever the ties of economic thinking from the "circular economy" and focus attention on the connection of economy to the Earth. Yet the fact that the economy functions as "circular"—or to use my term, "supra-material"—must be part of the materialism that G-R advocates. The duality between the economy and Earth is not merely an oversight in economic thought: it is an economic reality that must be accounted for if the ideas of the earthly philosophers are to fully rise to the historical moment.

The main current of EE has gone in the direction of G-R's student, Herman Daly. Daly brought the revolutionary thinking about limits embodied in the laws of thermodynamics to a wider audience. He reinforced a simple materialism (the economy is connected to the Earth) and married it to a neoclassical/Keynesian framework where the market is thought to work but needs government management to iron out its formidable problematic edges that extend to growth and distribution. In very simplistic terms, his thesis is laid out like this (and I use his terminology): humans have moved from "an empty world to a full world" and can no longer ignore the problem of growth and the need for limits. Economic growth has become "uneconomic"—the benefits from it are outweighed by the costs.[23] Daly proposes a steady-state economy where growth is limited by policy and inequality is moderated through redistribution. In the end his approach is prescriptive—put a physical cap on throughput—the materials and wastes flowing through the system—and redistribute income—and then let markets with a small *m* adjust.[24]

Exactly what the difference between a market with a small *m* and one with a big *M* is we are never quite sure, but this is what Daly says: "The Market with a capital 'M' is indeed a poor master and should be demoted to 'markets' with a small 'm' which can be good servants."[25] In my mind that's a bit like saying the cells in your body are good at what they do, but we just don't like the way they are put together to form the whole that is you. Daly is essentially saying that the market economy can and needs to be managed because it is good at registering information about what to produce and how to produce it once boundaries have been set. It seems that his taxonomy of markets might be a theoretical concession to his own distaste for planning.

In the end Daly leaves hanging questions that ought to be on everyone's mind among those that care about the Earth and see the necessity of ending growth. Can we manage capitalism not to grow? Or put another way—why have capitalism as your economic system if you don't want growth or inequality, and you can't abide ecological degradation, since on the face of it these appear to be the most salient features of the system. It might be important to have a better handle on the disease in order to come up with a cure, otherwise policy will always be swimming against the current of economic forces—and in the end, it seems the economic forces will prevail.[26] A more incisive analysis of the system—that focuses on its dynamic of expansion and generation of surplus and the contradictions therein, the profound extent of interdependence, and the duality between economy and Earth embodied in this system within a system where it is both material and supra-material would prove beneficial in understanding the impossibility of managing the system to be something it isn't.

I am sympathetic to Daly's inclination that it is very difficult to foundationally alter this economic system. Dismantling the tapestry of capitalism is no simple matter, especially if it is to be done in a time frame that allows us to stave off the worst effects of climate change and halt the sixth mass extinction. The truth is, we are between a rock and a hard place. The history of the last ten thousand years tells us that the trajectory of the economic superorganism has not been altered with inventiveness and other cultural levers of change. In fact, that history has landed us here with an economic system that is the most exaggerated form of the economic superorganism that we can imagine. A massive system both disconnected and connected to Earth. A system that we are now required to both grow and degrow.

It is strategic as opposed to sufficient to answer the challenge of duality and paradox (that the system is both material and supra-material, that we are now required to both grow and degrow it), with the idea that capitalism can be managed not to grow and made to account for the value of "natural capital" and "ecosystem services"—in other words, it can be reconnected to the Earth with these tools while we leave the system within a system in place. This is what is euphemistically referred to in the literature as development without growth, and it relies heavily on market signals and motivations as well as the ability to enact policies that will run contrary to the impulse of the system.[27]

EE has convinced many people that endless growth on a limited planet is not possible, and it has increased the awareness of the biophysical foundations of the economic system. It is a revolutionary message to recognize the need for limits in an economy such as ours and to acknowledge that the economy is connected to the Earth; however, a revolutionary message is not a recipe for revolutionary change. The economy continues its war with Earth. Consider where we stand according to David Attenborough: "Today, we ourselves, together with the livestock we rear for food, constitute 96% of the mass of all mammals on the planet. Only 4% is everything else—from elephants to badgers, from moose to monkeys. And 70% of all birds alive at this moment are poultry—mostly chickens for us to eat. We are destroying biodiversity, the very characteristic that until recently enabled the natural world to flourish so abundantly."[28] And on the other side of the extinction crisis is the apogee of the economic superorganism with its extreme wealth and extreme poverty. Presently 9–10 percent of the global population remains in extreme poverty. These numbers bring the challenge of the historical moment into sharp focus. Clearly EE remains a work in progress.[29]

Ecological Political Economy

It isn't surprising that a more radical group of earthly philosophers interested in the war between economy and Earth tapped more fully into a systematic and critical analysis of capitalism. They have attempted a more incisive analysis of the present structure and dynamic of the economic superorganism. They step into a void left by EE. The list of those who interpret the present war between economy and Earth as a war between capitalism and Earth is by now very long.[30] As with my brief elaboration of EE, my intention is not to provide an extensive review of the origins or breadth of this school of thought nor to elaborate on the internecine conflicts within this group of thinkers. These earthly philosophers believe that a foundational alteration in the economic system must occur in order to change the dynamic of surplus, expansion and stagnation, inequality, and ecological degradation. And all adopt a similar framework for understanding capitalism; that is, it is a system of exploitation (of both humans and nature), inequality, capital accumulation, and crisis. There are now many organizations and movements that adopt this framework to one degree or another—the Degrowth Movement is one of the many that come to mind. The intellectual foundation of ecological political economy is found in Marx and Engels and the twentieth-century interpretation of their work by Paul Baran and Paul Sweezy. It also extends to the more radical interpretations of Keynes (post-Keynesian) and to a lesser extent in the ideas of Veblen and other evolutionary economists.

There is no question this group of thinkers has a more critical and expansive approach for understanding the capitalist system and a more systematic analysis of the exploitation of humans and the Earth found therein. They carry materialism out of its superficial approach utilized by EE. But while ecological political economy (EPE) utilizes historical materialism to see the movement of economic systems over time, and to expose the foundational force and dynamic of capitalism, this group does not carry this methodology much beyond capitalism. As this book has illustrated, a more expansive materialism that highlights the foundational shift that occurred with the agricultural revolution sheds light on the tension between humans and the more-than-human world that extends beyond the imperative of capital accumulation and reaches into the complexity of economic order and its formation where evolutionary forces, system dynamics, and the complex dialectical interplay of humans and Earth are fully registered. Here more humility in the face of the present challenge, and greater reflection on the place of humans on Earth might and should be nurtured. A deeper materialism carries us back in our evolutionary history where we register the full challenge of the historical moment.

Let me give two examples of the inclinations of EPE. Again, I emphasize that this discussion is not intended to be exhaustive. I highlight the work of Jason W. Moore (and Raj Patel) and John Bellamy Foster and Paul Burkett. Jason W. Moore proposes renaming our present period the "Capitalocene" (as opposed to the Anthropocene) in the spirit of focusing attention on what he sees as the source of the problem, and in an effort to see capitalism as a "world ecology." Moore tells us "human organizations are environment-making processes," and he employs historical and dialectical materialism in his analysis when he claims an obvious truth: "Species make environments; environments make species."[31] Capitalism is its own form of this interplay and forms its own "world-ecology of power, capital and nature, dependent on finding and co-producing Cheap Natures."[32] With this Moore moves the discussion of the process of exploitation and capital accumulation beyond the narrow confines of wage labor to include the natural world, slavery—all forms of what he calls "cheap natures" on which capitalism depends.[33] Moore places no emphasis on a deeper materialism that recognizes the formation and power of the economic superorganism.

Moore's (along with coauthor Raj Patel's) discounting of the longer arc of human history becomes clear in this passage: "Hunting large mammals to extinction is one thing, but the speed and scale of destruction today can't be extrapolated from the activities of our knuckle-dragging forebears."[34] Would these knuckle-dragging forebearers be the same *Homo sapiens* who painted frescos on the walls of the Lascaux Cave twenty-thousand years ago and did the same thing that long ago in other places? Human prehistory was not a time where "men lived in mental and social twilight, waiting, straining to become fully human," and to assume otherwise is to be mostly ignorant of human prehistory.[35] Prehistoric humans were fully human in all that that implies. They were intelligent, observant, cultural, and innovative. They were also mostly nonexpansionary, minimalist, and embedded in the rhythm and dynamic of the more-than-human world in economic life, and they were mostly not in "overshoot" of the ecologies through which they roamed. They provide a good outline of a different economic context, a different economic system, one that forged a different relationship between humans and Earth. One can argue they were more us than we are. More importantly they provide a pivotal understanding of how dramatically humans were altered with the advent of grain agriculture thereby nurturing a more foundational curiosity about the formation of economic systems and a more central focus on the contextual complexity of the human relationship to Earth embodied in economic order—one that extends beyond the process of capital accumulation.

Moore is captured by the endgame of the economic superorganism. It's hard not to be. There is no question that a problematic world ecology of capitalism cannot be sustained indefinitely by any definition of that latter term. And there is no question we need "to forge a different ontology of nature, humanity, and justice—one that asks not merely how to redistribute wealth, but how to remake our place in nature in a way that promises emancipation for all life," as Moore suggests.[36] Yet as long as the focus of the problematic relationship of humans and Earth remains solely tied to capitalism the complexity and challenge of a deep ecology of economic order can never be fully engaged. And it is important to understand that capitalism as world ecology with its use of "cheap natures" is not at an end yet. There is a tendency among those critical of it to prematurely announce its postmortem, but the long arc of history tells us that the resilience of the economic superorganism is not something to discount. This bent in human history has been with us a long time. The truth is that capitalism hasn't run dry of its possibilities for accumulation and expansion and extending the duality between humans and Earth especially given our capacity to manage and accommodate the system, and to think we're changing something we aren't changing. There is a certain misconception in believing that the ecological crisis has become so bad there is now nowhere for capitalism to turn to shift costs to bolster capital accumulation as Moore suggests. Unfortunately, there is still room to further erode ecological systems, exploit humanity, and continue species extinction. What if our historical moment is not about immediate collapse—not collapse of capitalism and not collapse of Earth—instead it is about languishing in the twilight of capitalism, this apogee of the economic superorganism as we take that final step to an irrevocable and final chiseling into *Homo sapiens agriculturii* and the deafening silence of the sixth mass extinction.

John Bellamy Foster and Paul Burkett exemplify a slightly different bent in EPE. They similarly focus their attention on capitalism and elaborate all of the ways that Marx and Engels highlighted concerns about the environment.[37] Theirs is an agenda to retrieve the reputation of Marx on the ecological front. I can appreciate their inclination, especially given the high discount rate that EE has applied to Marx and his ideas. But in highlighting the ecological bent in Marx (and Engels) Foster and others seem to miss the importance of the fact that ecological concerns were not the central focus of their analysis. As I have stated, Marx and Engels were mostly in the camp of the unearthly philosophers, and there was a reason for that. Capitalism does function as if removed from the Earth. It was easy to concentrate on class tensions, the exploitation of workers, and the internal contradictions of this system without bringing biophysical connections into the mix. Remember, the system embodies duality by being two things—both supra-material and material.

The thorough probing of Marx and Engels by Bellamy Foster and others reveals that Marx and Engels were nevertheless beginning to develop foundational ideas about the dialectical interplay of humans and Earth. Foster and Burkett describe what they understand about Marx and Engels and their approach: "The relation between the organic body of a human being and the inorganic world is one that is conditioned by the subsistence needs of humans beings and their capacity through social labour to transform the 'external' conditions of nature into a means of satisfying these needs. . . . Marx thus attempted to describe the material interconnections and dialectical interchanges associated with the fact that human species-being . . . finds its objective, natural basis outside of itself, in the conditioned, objective nature of its existence."[38] Yet, the significance of social labor and its relationship to Earth fundamentally shifted when humans became engaged with grain agriculture. This was a constitutionally different economic world than what had been in place for somewhere between two hundred thousand and three hundred thousand years before the dawn of agriculture. It was a break, a new whole, and not a continuum. It was a novel play on human sociality. Neither Marx and Engels nor Foster and Burkett emphasize the importance of this change in recalibrating the relationship of the economic system to Earth, nor the play on the evolution of human sociality that it entailed: the former because their understanding of human prehistory and evolution was limited, and both the former and latter because the longer arc of history was discounted in order to focus on capitalism.[39]

An appreciation for human prehistory and the significance of the agricultural revolution as an altered mode of production reveals a great deal about the complexity of social labor, the objective conditions of material existence, and the formation of economic systems and the context they establish for the human/Earth relationship. Discounting the importance of this history means missing a foundational turning point in the "conditioned objective nature" of human existence and its formation. Foster and Burkett recognize the deep materialism that Marx and Engels were being led to, a materialism that was leading them to the heart of human evolution and the human relation to Earth.[40] There is no question that humans emerged as uniquely social as exemplified by their capacity for culture. But the question is not that they become uniquely social but how that sociality became expressed in an economic system and in relation to Earth.

Unfortunately, Foster and Burkett have not extended the deep ecological inclination of Marx and Engels in light of our present understanding of human evolution and human history and prehistory. It is one thing to be uniquely social; it is another to be collectively conjoined in a food-producing system with profound positive feedback loops. In fact, one might claim (and I have clearly argued) that human sociality was hijacked by the agricultural system creating a system of duality between humans and Earth that had not previously existed. An agricultural system establishes a relationship of domination and separation, and begins a time when humans ceased "to listen to a million secret tongues."[41] Of course, capitalism takes this inclination and turns it into the most perverse expression of human sociality in relation to Earth and in relation to other humans—in exploitation, in assembly line production, in the alienation of humans from what they produce, in the fetishism of commodities, in the metabolic rift of the economic system and the Earth, in the extinction crisis.[42]

Foster and Burkett certainly recognize that "the basic answer to the global environmental problem is at all times the same: the struggle to recreate a balance in our relations to the earth—before the earth system . . . creates a balance of its own—one outside the contours of what constitutes a safe operating space for humanity."[43] A more expansive view of our "environmental problem" carved out of the deep arc of history might lead us instead to ask whether we have the capacity to recreate a balance in economic life that orchestrates a safe operating space for the more-than-human world?

Ecological political economy has contributed to the toolbox of the earthly philosophers, but it seems necessary to take their historical methodology and carry it back into the deep time of *Homo sapiens*. Humans came to the Holocene, a smart, self-conscious species with a well-developed capacity for sociality. But the emerging system of grain agriculture with its powerful feedback loops played on human potentiality in a way that created an economic superorganism not that different from the ones in ants and termites that engaged cultivation. Once a powerful integrated energy system dynamic took hold, it began to have a life of its own—a unique and formidable force in the matrix of evolution. It captured and changed *Homo sapiens* and reconfigured their individual and collective relationship with Earth. It also reconfigured their relationship to each other. The rest is history, we might say. Despite culture, intelligence, and the rest of our unique capacities we appear to have been carried along by a powerful system dynamic that has reached its apogee with global capitalism. Again, we must ask, what of its twilight? How do we navigate its twilight?

The great scholar of economic thought, Eric Roll, tells us that "the economic structure of any given epoch and the changes that it undergoes are major influences on economic thinking."[44] Surely the landscape of duality, paradox, ecological decay, extinction, growth and stagnation, and poverty are the hallmark of this economic epoch and expansively understanding this reality is the challenge for the earthly philosophers. An appreciation for deep history and a methodology that embraces and encourages it are essential tools of this craft. In this matter of Earth and economy looking forward actually means looking back and taking the challenge of history seriously. And if the challenge of history is taken seriously, economic thinking must bring the proposition that Earth changes humans and humans change Earth to its full historical meaning, giving deeper insight into economic system formation and the way humans are contextualized by them.[45]

The economic scribblers of all persuasions might keep in mind how they will be perceived two hundred years from now. Their ideas will not be evaluated on the basis of whether they captured the business cycle, inflation, growth of GDP, or the rise and fall of asset markets. Their scribbling will be evaluated on the basis of whether they provide insight into the challenge of the historical moment; that is, the challenge of confronting the war between economy and Earth and dismantling duality. Expansively framed ideas can help to refine the levers of change at our disposal and allow us to think more incisively about their purpose. How with a human population of eight billion people and an economic system in a dialectical tension requiring both growth and limits at the same time do we reembed economic life in the rhythm and dynamic of the more-than-human world? How do we create an economy where a finely tuned human ecology results; that is, where humans once again take their place as one of many? Do we have the capacity to do this? What is the power of the economic superorganism? Perhaps more importantly these are the important questions for the economic scribblers—not whether GDP will grow or shrink next quarter. The earthly philosophers have begun this project, but it remains unfinished.

Epilogue

Languishing in the Twilight of the Apogee

Contemplation

There is a question that has shadowed me throughout this inquiry: how can we be of this Earth and yet have moved so far in the direction of disrupting its well-established cycles and self-willed otherness? The process of evolution is all about creating something distinct out of what's already there, but what do we make of something as antithetical to ecological life as we have become through the legacy of our Holocene economic formation, which is to say our Holocene evolution. We observe the refined ecologies and coevolutionary outcomes of life on Earth—orchids and their insect pollinators, the coevolved beaks of birds, the life cycle of the monarch and its migration—magic. And yet here we stand, evolved to what we have become, systematically obliterating these temporary perfections.

It is perhaps important to keep in mind that perfection itself is accident (and actually impossible to prove), and evolution doesn't see ahead. Evolutionary processes have no ultimate goal, no teleological purpose. Along the way, processes of evolution reliably give outcomes that aren't perfect or, more importantly, viable in the long run. Think of all the species that inhabited the Earth at one time or another—moments of perfection (or not) in their time. Yet it is true that humans are placed in a unique and dramatic situation in the context of evolution. As the agricultural revolution demonstrates the human capacity for culture and human intelligence allowed for quick and intentional adaptation—expanding and accelerating the system dynamic, the new whole, that was taking form with agriculture. The truth is we became *Homo sapiens agriculturii* without measurably altering our DNA; this was our evolutionary dance. The process of elaboration around cultivation was a much longer process of mutation and selection for our insect counterparts.

We are confronted with a complexity of our own evolution that we find difficult to navigate. We have come to believe that the power of adaptation and change is entirely in our hands—after all, we have evolved to be an intelligent, cooperative, cultural, inventive species, and these qualities presumably give us control over our destiny. We place stock in our power to call the shots through these capabilities and their manifestation in specific levers of change. And yet the evolutionary message of the agricultural revolution is that we are subject to the same forces and system dynamics as other species, and our unique qualities merely shortened the temporality of the evolutionary process that resulted in the formation of the economic superorganism. The truth seems to be that our unique capacities have been accommodating to the agricultural system dynamic both in its formation and its subsequent unfolding. These capacities have yet to run counter to it. The question that stands before us now is whether we have the power and will to override the structure and dynamic of the present iteration of the economic superorganism. Which is to ask: do we have the power to influence our present evolutionary trajectory?

We find ourselves two things at once—we are an economic superorganism, but we are also the genetic result of our Pleistocene evolution. We are both *Homo sapiens sapiens* where we evolved as one of many species in a world mostly not of us. And we are *Homo sapiens agriculturii,* a collective solipsistic species adhering to a particularly problematic entropic path. It is difficult to calibrate what the loss of the Pleistocene environment means for us and what it will mean to continue on this evolutionary trajectory. It is pretty clear what it means for the wild more-than-human world. What is the difference between being an economic superorganism and a species that is embedded in the rhythm and dynamic of the more-than-human world?

You might say we now reside in a liminal space—somewhere between determinism and intentionality—between what we can do and what we are caught up in, between an inclination to take our place as one of many species, and the forward march of the economic superorganism and the sixth mass extinction. Given the difficulty in navigating this space we have learned to live with its contradictions and hedge its challenge. In a sense we languish both individually and collectively in the paradox of this historical moment. Individually we confront the paradox of the historical moment as tension in our daily lives where economic survival and economic security demand adherence to the economic system but rapprochement with Earth demands the opposite. In the realm of system change the tension appears in the impossible challenge of reducing inequality, poverty, and stagnation in a system that reliably generates them, at the same time we are required to reduce the material manifestations of economic life.

We are thrown into the realm of fanciful thinking, adhering to age-old ideological ruts and our power to change things, believing there are no diminishing returns to efficiency and our inventive capacities, and conflating real and spurious rapprochement with the Earth. (As a reminder, real rapprochement with Earth is the challenge of moving to an economic system where we humans take their place as one of many species on Earth.) Perhaps we need more reverence for the power of this moment and for what we are caught up in. This is a moment that demands we contemplate our evolutionary history, our place on Earth, and the long arc of history. It is a moment that challenges us to think more critically about our effectiveness to alter our circumstances and remap the context of our collective economic lives in relation to Earth.

In more stark economistic terms the challenge we confront is that we are required to move a world of almost eight billion people, almost all involved in an economic system with tremendous inequality and poverty, a clear imperative to expand and a chronic tendency to stagnate, to some real as opposed to spurious rapprochement with Earth. We are required to stop the impact of this behemoth on "the global climate, ocean chemistry, the cryosphere, the nitrogen cycle" and retain the "abundance, diversity, and distribution of fauna and flora" at the same time we reduce poverty and inequality.[1] On the face of it, this seems to be impossible and yet is an accurate description of our challenge. If we are serious about retaining the "abundance, diversity, and distribution of fauna and flora" then we are talking about levers of change that run counter to the impulse of the economic superorganism.

In the very long arc of Earth time what we do or don't do doesn't really matter. Yet we reside in the arc of the next five hundred years—the end run of an arc of history that has been with us for ten thousand years. If we are to stop ourselves from crossing a great divide; that is, if we are to stop the sixth mass extinction we are forced to readjust the aperture of the focus of our levers of change. We clearly need more depth of field. Unfortunately, it appears that our levers of change are mostly employed without depth of field and in the service of our present trajectory. In the end we use them to hedge the challenge of the twilight of the apogee of the economic superorganism. Let there be no question that challenge is the challenge of limits and real rapprochement with Earth. I offer two examples of how levers of change lack the force and framing to rise to the historical moment.

The Conservation Compromise

In an economic world engaged in the sixth mass extinction there is nothing more important than the conservation of the wild. Nothing. It is a line in the sand and a clear expression of the recognition of limits; a significant form of resistance against domestication and the power of the economic superorganism.[2] I interpret the conservation of the wild rather broadly here to create space (literal space) where the right of existence of self-willed otherness over human use and need is prioritized. In the United States, the levers of change used to execute this mission are many but would include the Wilderness Act of 1964, the Endangered Species Act, and the establishment of National Parks. They might also be extended to include the resistance against the placement of wind farms and pipelines, and more globally the expansion of agriculture. This list is clearly not exhaustive.

The debate over "wilderness" and its place in conservation exemplifies the tendency to hedge what needs to be done in the face of the sixth mass extinction. Wilderness and its protection have become contentious issues in conservation circles. To some extent that contention revolves around a foundational disagreement surrounding the relationship between humans and the more-than-human world—specifically, whether that relationship should be viewed from the perspective of duality or dialectics. This tension between duality and dialectics is somewhat contrived. There is a powerful dialectical interplay between humans and Earth—but that doesn't negate the presence of the powerful duality that took form when humans began the cultivation of annual grains. Dialectical thinking must be brought into the long arc of history where it recognizes qualitative change. A powerful whole was formed with agriculture that separated humans from Earth in a way they previously had not been separated. In fact, without the recognition of this duality—its structure, dynamic, and evolutionary significance—conservation remains in the shadow of confusion, a diminished lever of change.

Those who see wilderness as something that stands apart from humans and must be protected are accused of ignoring the dialectical relationship between humans and nature. This is a powerful current in the criticism of "deep ecology." The historian William Cronon laid the foundation of the approach that treats dialectical interplay and duality as in opposition. He stated in no uncertain terms, "To the extent that we celebrate wilderness as the measure with which we judge civilization, we reproduce the dualism that sets humanity and nature at opposite poles. We thereby leave ourselves little hope of discovering what an ethical, sustainable, honorable human place in nature might actually look like." Cronon more directly takes on the deep ecology movement claiming, "When they express, for instance, the popular notion that our environmental problems began with the invention of agriculture, they push the human fall from natural grace so far back into the past that all of civilized history becomes a tale of ecological declension."[3] Unfortunately, history tells us there is a clear connection between ecological declension and the agricultural revolution. More importantly, if one discounts the importance of the agricultural revolution one has missed one of the most significant alterations in human evolutionary, entropic, and economic history: an alteration that foundationally recalibrated the human relationship to the more-than-human world. This change set humans on the path that landed them here in the vortex of this system within a system.

And, as an aside, we might keep in mind that ten thousand years ago human global population stood at somewhere between six and ten million people. The Earth was, in fact, a place, that functioned mostly without human culture. There was, in fact, an Earth composed almost entirely of self-willed otherness—call it untrammeled wilderness. And where humans did reside, their interaction with the more-than-human world was distinctively different than what came after settled agriculture. Humans are, if nothing else, contextual. Humans that lived before settled agriculture had an economic structure and dynamic to material life that was mostly embedded in the rhythm and dynamic of the more-than-human world. Put in the language of domination, it was the impulse of the more-than-human world that directed humans and not the other way around—and most importantly, humans were not caught up in an economic system that created a structural duality between them and Earth. It is, in fact, a stark reality that in ten thousand years we have gone from a world where there was no culture/nature dichotomy and lots of untrammeled wilderness to a world of almost eight billion humans and the sixth mass extinction. It is this particular arc of history that these pages have attempted to acknowledge and elaborate. And it is the arc of this particular history that we must now confront in the face of the sixth mass extinction and the apogee of the economic superorganism. An uncompromising approach to conservation of the wild is essential to the historical moment.

Yet Cronon set the stage for critiques of deep ecology, and in the decades that followed many progressive thinkers in the field of conservation took his lead. The conservation mission has been altered as a result. The revised approach to conservation initiated under the leadership of Peter Kareiva, formerly of the Nature Conservancy, is one example. More recently, Rebecca Solnit's critique of the history of the Sierra Club, and especially the framing of its mission by John Muir, is another example. Their approach to conservation emphasizes the need for social justice and the necessity of establishing "workable" ecological relationships between humans and the world around them. No one argues against creating workable ecological relationships between humans and Earth nor social justice, but this can easily take the form of hedging the necessity of limits and furthering the complete domestication of the planet.

Peter Kareiva, Michelle Marvier, and Rober Lalasz write in their article "Conservation in the Anthropocene": "Conservation's continuing focus upon preserving islands of Holocene ecosystems in the age of the Anthropocene is both anachronistic and counterproductive. . . . Conservation must demonstrate how the fates of nature and of people are deeply intertwined—and then offer new strategies for promoting the health and prosperity of both." In the effort to employ this new twenty-first-century conservation Kareiva, Marvier and Lalasz claim that "conservationists will have to jettison their idealized notions of nature, parks, and wilderness."[4] They describe the more progressive form of conservation for the twenty-first century: "The bigger questions for 21st century conservation regard what we will do with . . . the working landscapes, the urban ecosystems, the fisheries and tree plantations, the vast swaths of agricultural monocultures, and the growing expanses of marginal agricultural lands and second growth forests that, as agriculture and forestry become more productive and intensive, are already returning to something that may not be wilderness, but is of conservation value, nonetheless."[5] Unfortunately, it is this larger goal that has come to dominate the conservation efforts of the Nature Conservancy and it has, in fact, jettisoned its commitment to preservation of the wild.

No one doubts the importance of the type of conservation to which Kareiva and his ilk aspire—they attempt to create a more sustainable relationship between humans and Earth involving human material activity. But to "jettison" wilderness preservation in its stead is simply to miss the long arc of history as well as the significance of the historical moment. To make matters worse, this approach to conservation demonstrates an astounding ignorance (at least on the part of Kareiva and his coauthors) of the present iteration of the economic superorganism: global capitalism. It is a system reliably creating poverty, inequality, stagnation, growth, and waste. To believe we can create a new and improved conservation around this foundation is its own expression of illusion. It's like mopping up the decks as the *Titanic* sinks. It isn't acceptable to substitute an accurate assessment of the economic system and its present paradox and challenge, with some naïve belief that conservation can go about creating harmony in a sea of madness.

More recently Rebecca Solnit formulated a critique of John Muir and his prejudices using the same framework as that established by Cronon (and used by Kareiva and others). Solnit states that Muir "carried the prejudices of most white people in his time." In an atmosphere of heightened awareness of racism her critique resonates, but it is the way she connects it to the idea of wilderness that is the focus here. Solnit claims that the idea of wilderness runs counter to US history because when Europeans arrived on the shores of what is now the United States they fully understood "that they were entering someone else's homeland, that these places were fully inhabited. . . . They knew they were invaders, partly because they fought to dispossess Native Americans." She continues: "The idea that much of this continent was wild in the old sense of untouched or uninfluenced by human beings" was erroneous.[6] That is debatable, although it is not debatable that there were Native Americans dispersed across the continent.

In Solnit's framing, the environmental movement needs to understand "that there is no inevitable nature-culture dichotomy and that the example of people living on the land—or of many peoples living on many kinds of land, from the Arctic to the Everglades—without devastating It has always been here."[7] This is not untrue—it just isn't the whole story. The relationship between humans and nature that began with grain agriculture is dramatic expression of dichotomy and dualism, and this dualism became the dominant economic and cultural impulse in the ensuing ten thousand years. Can we learn something about living sustainably from people who have done a better job? Of course we can, but we can't ignore the arc of the legacy of grain agriculture. This is precisely the arc that informed John Muir's approach to wilderness and the need to protect it.

Protecting the wild by acknowledging its existence is not antithetical to ecological living nor is it anachronistic; it is essential on a planet of almost eight billion people and with an economic system such as ours. It is a line in the sand that speaks to the necessity of limits. Real limits, line-in-the-sand limits, against domestication limits. It speaks to the fact that the human presence on Earth is too big, and it recognizes, without equivocation, the presence of self-willed otherness and its right to exist. It is a necessity in the end game of duality. Preservation of the wild as apart from humans does not negate the fact that we need to find an ecological balance with Earth in the securing of our day-to-day material life, but the parameters of that ecological balance must be set within the context of the rights of self-willed otherness especially when ecological balance in a system such as ours remains elusive.

I do not engage a discussion of Muir's racism, and I fully recognize that Native Americans were dispossessed of their land in the settling of America and through the various designations of wilderness. But I think it's important to recognize the impulse that Muir was reacting against in his vision of untrammeled nature. He was recognizing the speed and direction of an economic system that had been in play for a long time and was reaching its full exponential flight during the nineteenth century. It was a scorched earth dynamic and still is. Muir's elevation of the natural world (sans people) must be understood against the backdrop of this dynamic. He understood that without awareness and intentionality we might be left with little of that world.

Lest you have trouble imagining the pace of change that has taken exponential flight over the past two centuries let me refer you to Thomas Jefferson's first inaugural address where he commented on the attributes of the nation. He said that one of those attributes was that we "possessed a chosen country, with room enough for our descendants to the hundredth and thousandth generation."[8] If you actually do the math on his proposition what you find is that he believed there was enough land for an agrarian nation to expand for somewhere between 2000 and 20,000 years. Yet we were filled up in this respect within 100 years, and within 250 years we have landed here. Not all of his miscalculation owes to the arid interior west; indeed his was a miscalculation of the movement of the economic system. I'll grant him the context of his preindustrial mindset—but still.

It isn't terribly useful to say there is no inevitable nature-culture dichotomy when what we confront is precisely what appears to be an inevitable nature/culture dichotomy. In the last ten thousand years, a profound and dominant nature/culture dichotomy emerged, and it is the power and force of this duality that we now confront. The power of culture broadly construed (our ability to work together, our capacity for inventiveness, our intelligence) has not altered the forward march of what began with agriculture. Wes Jackson tells us humans became "a species out of context" with agriculture but it might be more accurate to say humans became a species of a different context. It is the form "entropic indeterminateness" has taken for *Homo sapiens* in the past ten thousand years, and it is an evolutionary divide that we now confront.[9]

If examining the racist history of the founder of the Sierra Club leads to the conclusion either that there is no inevitable nature/culture dichotomy or that the core of the Wilderness Act of 1964—the focus on setting aside untrammeled wilderness—is somehow misguided, then I would hold that the Sierra Club is on the same dangerous mission creep as the Nature Conservancy. It is swimming with the current of the historical moment rather than countering it. To acknowledge that Native Americans had balanced relationships with the Earth in areas that we came to consider wilderness is fair to point out. To say that humans change the Earth and the Earth changes humans and what we need to do is figure out some more culturally workable relationship is true. But the designation of wilderness is not a misunderstanding of history or an erroneous interpretation of the nature-culture relationship, it is a recognition of the long arc of history and the reality of a most pernicious nature/culture dichotomy. If conservation can't identify a clear designation of limits and the history that stands behind it, then I am correct: the levers of change at our disposal merely accommodate the forward march of a problematic system.

Solnit claims that we can't "truly protect a place by setting it apart."[10] It is true that we have ended up with islands of wildness amidst an irrevocably domesticated world. Just hike in the Tetons and look westward into Idaho if you don't believe this is true.[11] If we don't ultimately figure out how to effectively alter the structure and dynamic of the trajectory we find ourselves on, we will lose our fight to halt the sixth mass extinction, wilderness designation or no wilderness designation. Yet the designation of wilderness and its protection is an essential part of moving through this twilight as well as internalizing the importance of real rapprochement with Earth. Protecting and setting aside wilderness need not be jettisoned but instead pursued with an evangelistic earnestness.

Technological Optimism and the Entrepreneurial Spirit

There is no lever of change like technology, and there is no ideology that is more powerful than the belief that humans will solve their current problems in their relationship with Earth through their capacity for inventiveness and innovation. Technology should be understood as knowledge, understanding, know-how, and the manifestation of knowledge in ways of doing things (technics). The American economist Thorstein Veblen referred to it as the "instinct of workmanship." It is the result of the dialectical interplay of humans and Earth in the sense that out of this interplay humans evolved to be an intelligent, inventive, and cultural species.

Knowledge and technology are cumulative for humans, and they may be implemented consciously and with a purpose in mind; that is, to solve a particular problem. Humans may overcome temporary problems of scarcity and ecological limits and open up new vistas of abundance through technological innovation, but the desire for greater abundance is not necessary in order for the knowledge base to expand. Think of the three-hundred-thousand-year period of *Homo sapiens* living as hunter-gatherers in an economic system that was contained, minimalistic, and nonexpansionary. This is a remarkable history of adaptation to dramatic climate changes within a stable and nonexpansionary economic system. It teaches us that technology is gauged in relation to the system within which it functions. This is important to think about because we are now captured in an extremely expansionary system, and yet we do not need technology to accommodate this impulse but instead to move against its current.

Technology has certainly augmented the system (especially its expansionary dynamic) over the past ten thousand years. In other words, it has been accommodating to the structure and dynamic of the system often to the detriment of the longer-term stability of the system at play. Think about all the ecological problems that emanated from the use of irrigation: greater population growth, more soil erosion, deforestation, salination of the soil, among others. Yet our perception of our intelligence and technological abilities reinforces the sense that we have the power to dictate our destiny through our ability to direct our innovations to quickly address problems. History tells us otherwise; it tells us we have a history of a powerful system dictating the direction of technological innovation. In our current economic environment our technological abilities are associated with the entrepreneurial spirit, and this duo carries great ideological force connecting human capabilities for innovation with economic liberalism.

There are many who are believers in the potentiality of human inventiveness and the economic system in which it is nudged. Bill Gates and Steven Pinker are good examples of the parishioners of this particular church, but so are the Green New Dealers who want jobs and economic growth but also the ability to confront climate change. Technology is the thing that proved Malthus wrong (in the short run) about his proposition that population growth would outstrip food production, reinforcing the belief among some that infinite growth on a finite planet is possible. We actually believe the second law of thermodynamics is negated by human inventiveness and that colonies on Mars are a viable option for the future of humankind.[12]

The mass production technology emerging out of capitalism and the industrial revolution provides an excellent example of the way technology accommodates the system at hand. The first thing to understand is that mass production technology increased the productivity of the system exponentially. It allowed capitalists to reach economies of scale and lower per unit costs. But it is also important to realize that it reduced the power of the worker vis-à-vis the capitalist, thereby giving the capitalist more power to lower the indeterminacy between what they purchased when they hired a worker and the work they actually got out of them. Its purpose was not simply to increase productivity but to deal with the labor problem (to hold down wages and better control workers). In this process individual workers were further reduced to cogs in a machine, destroying any creative interaction with the world around them in their work life. Workers were deskilled and easily replaced.

Yet mass production technology created significant problems for the system particularly in terms of the tendency for overproduction and the exacerbation of inequality. In other words, it set the system up for internal contradictions that were formidable. Just as irrigation helped to sow the seeds of the destruction of Middle Eastern civilizations, for example, so too mass production technology in nineteenth- and early twentieth-century capitalism helped to create more formidable system problems. In the end capitalism was rescued from itself by Keynesian policy, the welfare state, and by war and other forms of wasteful spending. But the system has never adjusted to the biophysical limits of this "solution."[13] This is clear as we confront climate change and the continual demand for more energy.

The Green New Deal—the idea that the government can help to push for the use of renewable energy that will provide enough energy for the economy to continue to expand, thereby providing full employment without sacrifice from the wealthiest—and at the same time keep us within a 1.5 degree temperature change—is the present salve to soothe our economic tension with earth. In the long run, it is a decoupling proposition that promises to defy the logic of energy return on energy invested.[14] It isn't clear that technological change exists that will allow for economic expansion sufficient to keep the wolves of stagnation from the door, provide the impoverished with a good life (but not requiring the haves to give up anything), and keep humanity from overstepping planetary boundaries sufficiently to allow for the right of all the species with which we share the planet to flourish.[15]

There is no question there are possibilities for increased energy efficiency and for greater relative use of renewable energy, but we need to ask ourselves what the upper limits to renewable energy are. What are the energy demands of the transition? Where does it lead us?[16] Here it pays to reflect on what G-R had to say: "Solar energy has an immense drawback in comparison with energy of terrestrial origin. The latter is available in a concentrated form. As a result, it enables us to obtain almost instantaneously enormous amounts of work, most of which could not even be obtained otherwise. By great contrast, the flow of solar energy comes to us with an extremely low intensity."[17] What this means is that solar energy, although infinite in one sense, is not infinite in another sense. It must be concentrated, transported, and stored. It is not a costless energy and environmental proposition to do so.

Just as the amount of nonconventional fossil fuel has a limit set by the energy return on energy invested, so does renewable energy. It requires energy to get it into a usable and consistently reliable form. Renewable energy will never get us to the energy return we had in the heyday of conventional oil and gas, nor present-day coal. Eventually expansion of the array of solar panels, wind turbines, smart grid, and adequate storage will reach an upper energy limit, and it will certainly be a lower net energy proposition than conventional fossil fuel. Of course, there are other considerations as well—do we honestly want a planet literally covered with photovoltaic cells and industrial wind farms? Communities are rising up to say no. And even if renewable energy didn't have an upper limit there are many other resources that do.

There is no question we are running out of time to stop the buildup of greenhouse gases sufficiently to maintain lives in the future that resemble the world we have known in the recent past. And there is no question that the problems caused by climate change will readjust the global economy in ways that are hard to predict—destabilizing an already unstable system. It's difficult to predict the total outcome in the short and long run—instability and economic decline in some areas but certainly economic opportunity in other areas. Instability always translates into human suffering so we can count on greater human suffering as well as the acceleration of mass extinction. Renewable energy is the present salve that saves us from facing the prospect of limits, and facing the paradox of our economic system. It is part of an infinite regress of profit and technological elaboration that takes us further from any possibility of a real détente with Earth.

We need to be clear that the promise of 100 percent renewables by 2050 and the Green New Deal are their own forms of business as usual (BAU). They are a way to avoid talking about limits—and the limits of the economic system—both in the short run and in the long run. There is no question that good Keynesian policy (industrial policy) levers can be used to disseminate green energy technology more rapidly than if it were left solely to the private sector. (Fossil fuel corporations have a high threshold for transition.) Yet we should understand that the whole premise of this ilk is to work with capitalism, not against it. As it presently stands the Green New Deal is not an attempt to deal with the necessity of limits; it is a technological promise that says we can avoid the prospect of limits.[18] It promises the haves will not have to give up anything in order for the have nots to have more access to energy; it promises sufficient jobs will be forthcoming; it promises there are no limits to economic expansion; and it promises that solving the problem of inequality need not be done within limits. It promises infinite energy into a future of infinite expansion of the economic system. Clearly, it is the same old wine in a new bottle. It reinforces Steven Pinker's claim that "energy channeled by knowledge is the elixir with which we stave off entropy, and advances in energy capture are advances in human destiny."[19] Maybe not.

Twilight

My focus has been to get under the etiology of the economic system in order to bring greater clarity to the problematic economic emergence of the last ten thousand years and its present iteration. It is clear we are captured by a powerful system, and our unique human potentialities (sociality, culture, intelligence, inventiveness) are perhaps players in a grander performance—drawn into the vortex of the economic superorganism. The longer arc of history teaches us something about how we might frame our present challenge. It is the challenge of limits in a system geared to expansion. It is the challenge of assuring a good life for eight billion people in the context of real rapprochement with Earth. It is the challenge of coming to terms with our own evolutionary history. What happens in the twilight of the apogee of this system will determine the future for humankind in a foundational way. Can we step back from this great divide and recede? Or are we destined to cross over? One thing is clear from this preliminary look at levers of change—they are often poorly framed and they usually play around the edges of foundational change and instead facilitate the forward march of the economic system. It is imperative to be clear about what it is we hope to accomplish when we engage them: at the moment, this is anything that hedges the need for limits and the need to end the sixth mass extinction.

It seems clear that the economic superorganism can be better or worse in the following sense: we can have more inequality or less inequality; we can have more pollution or less pollution; we can make a quicker or longer energy transition and have more warming or less warming. Doing things better is not the same as facing limits and engaging real rapprochement with Earth. It might simply mean it will take us a bit longer to get to the final act where the system has moved irrevocably to the complete annihilation of wild species and the project of 'humogenizing' the Earth.

The simple dictum that Daly left us stands—we can't have infinite growth on a limited planet, but in the twilight of the apogee this dictum is not enough. We need a more focused orientation. Do we want to substitute renewable energy for fossil fuel or work toward real rapprochement with Earth and confront the necessity of limits? Do we want to solve the problem of inequality within limits or solve it by expanding? Do we want to attempt to engineer a more harmonious relationship with Earth through our conservation efforts, or do we want to retain its wild impulse? These are not the same. Do we recognize the need to downsize the human population, or do we simply collapse the conversation about population into an infinite labyrinth of reasons why it's not really a problem? As we stand at this critical historical juncture we need to ask a simple question of our levers of change: what do we hope to attain when we use them?

Notes

Preface

1. The many books that Paul wrote include: *The Tender Carnivore and the Sacred Game* (New York: Scribner's, 1973); *Nature and Madness* (San Francisco: Sierra Club, 1982): *Coming Home to the Pleistocene*, ed. Florence R. Shepard (Washington, DC: Island/Shearwater, 1998).

2. Paul offered this observation on the matter: "Few prehistorians suppose that those earliest farmers and first domesticators east of the Mediterranean were conscious revolutionaries or even that changes were dramatic in a single lifetime . . . Yet by the time civilization began in the great city-states of Egypt and Mesopotamia, the tradesmen, bureaucrats, and tillers of the soil exceeded their hunter forebears in possessions and altered their surroundings—and were the creators and victims of new attitudes, expectations, and mythology." Shepard, *Nature and Madness*, 19.

3. Wes Jackson began the Land Institute in the 1970s with his first wife Dana. The Land Institute has grown into a major research institute for the development of perennial grains grown in polycultures. It has more recently expanded to include education and research with the goal to shift culture in the direction of a more ecological way of living on earth.

4. Wes Jackson, *Nature as Measure: The Selected Essays of Wes Jackson* (Berkeley, CA: CounterPoint, 2011), 4–5.

5. Obituary of Alfred Crosby, *New York Times*, April 4, 2018, national edition.

Introduction

1. When I refer to the agricultural revolution in what follows I am specifically referring to the cultivation of annual grains. I realize annual grains were not the only plants humans cultivated and that along with cultivation came the domestication of animals. I do not focus on these because I am interested in the story of economic formation and annual grains were the pivotal axis in that formation.

2. Many scholars recognize that agriculture was ecologically destructive, but the fundamental alteration in the trajectory of human evolution is not commonly noted.

3. The work of Jason W. Moore is illustrative of this trend. His orientation will be discussed in due course.

Chapter One

1. I refer to material order, economic order, and economic and material life interchangeably and use these terms to mean the production of material life, namely food, shelter, clothing, etc.

2. When I speak of the agricultural revolution in humans I refer specifically to the cultivation of annual grains.

3. Steven Pinker comes to mind. His book *Enlightenment Now* argues that we are on a trajectory of progress given our unique capacity for reason, our humanism, and our capacity for invention and innovation. My work stands in contradistinction to this claim.

4. The human inclination in this direction is unequivocal and universal. After the Holocene warming, grain agriculture begins independently in different parts of the world. The form taken by societies that engage in grain agriculture is all of the same ilk, regardless of their differences in culture.

5. These embellishments certainly gave humans the impression they were in control.

6. Among the few are myself and John Gowdy. See John Gowdy and Lisi Krall, "The Ultrasocial Origin of the Anthropocene," *Ecological Economics* 95 (2013): 137–47; John Gowdy and Lisi Krall, "Agriculture as a Major Evolution Transition to Human Ultrasociality," *Journal of Bioeconomics* 16 (2014): 179–202; John Gowdy and Lisi Krall, "The Economic Origins of Ultrasociality," *Behavioral and Brain Sciences* 39 (2016), e92; John Gowdy and Lisi Krall, "Disengaging from the Ultrasocial Economy" *Behavioral and Brain Sciences* 39 (2016): e119; Lisi Krall, "The Economic Legacy of the Holocene," *Ecological Citizen* 2 (2018), 67–76; Lisi Krall, "Reckoning Our Ultrasocial Past," *Trumpeter* 31, no. 2 (2015): 102–111.

7. Tim Flannery, "The Superior Civilization," *New York Review of Books*, February 26, 2009. Attine (aka *atta*) is the genera of the most sophisticated of the agricultural ants: the leafcutter ants.

8. Bert Hölldobler and Edward O. Wilson, *The Superorganism* (New York: W. W. Norton), 2009.

9. Edward O. Wilson, *The Social Conquest of the Earth* (New York: W. W. Norton), 2012.

10. The work I did with John Gowdy is one of the few attempts to do this.

11. Gowdy and I labeled humans (and insect cultivators) as ultrasocial, and we associated ultrasociality specifically with agriculture. I am inclined now to call

agricultural species *economic superorganisms* in order to stress and emphasize the integrity of this unique and universal system. My emphasis is that it is a new whole with standing in the matrix of evolution.

12. In evolutionary biology reproductive fitness is simply referred to as "fitness."

13. Edward O. Wilson, "One Giant Leap: How Insects Achieved Altruism and Colonial Life," *Bioscience* 58, no. 1 (2008), 17.

14. Hölldober and Wilson, *Superorganism*, 9.

15. Hölldobler and Wilson, *Superorganism*, 4.

16. Kin selection had greater currency in explaining altruism in insect superorganisms, but even here on closer look genetic relatedness does not fall into seamless patterns.

17. Wilson, *Social Conquest*, 181. David Sloan Wilson, one of the seminal thinkers in advancing group selection, acknowledges the importance of Lynn Margulis's proposal that the eukaryotic cell is "a symbiotic associate of bacteria." Margulis's proposition bolstered the case for the significance of group selection. The presence of higher-level organisms implies a dominance of the group (as opposed to the individual) in the evolutionary tension between the individual and the group. David Wilson tells us that "few biologists noticed that Margulis's theory, which involves between-group selection trumping within-group selection, was diametrically opposed to the dogma that within-group selection invariably trumps between-group selection." See David Wilson, "Multilevel Selection and Major Transitions," in *Evolution the Extended Synthesis*, ed. Massimo Pigliucci and Gerd Müller (Cambridge, MA: MIT Press, 2010), 88. A list of the notable evolutionary biologists, anthropologists, economists and philosophers involved in this discussion includes but is not limited to Lynn Margulis, John Maynard Smith and Eörs Szathmary, Samir Okasha, David Sloan Wilson, Edward O. Wilson, Massimo Pigluicci and Gerd Müller, Eva Jablonka, Marion Lamb, Kevin Laland, Peter Corning, Pete Richerson, Robert Boyd, Samuel Bowels, and Herbert Gintis.

18. Edward O. Wilson and David S.Wilson, "Rethinking the Theoretical Foundation of Sociobiology," *Quarterly Review of Biology* 82, no. 4 (2007), 328.

19. Wilson, *Social Conquest*, 243.

20. Samir Okasha, *Evolution and the Levels of Selection* (Oxford: Clarendon, 2006), 178.

21. It should be noted that there is a long tradition of exploring the collective as opposed to the individual among economists and other social scientists. This disposition has lost some of its currency in current economic discourse, which is so preoccupied with individual decision-making as the central focus of economic analysis. There is a tradition (e.g., Smith, Marx, Hayek) to explore the whole as something significant. And sociologists (e.g., George Herbert Mead and Emile Durkheim) have explored the dance between the individual and the group in detail. None of these luminaries has concentrated on collective ordering of species around agriculture per se, but this seems to be a logical place to explore the collective

given the importance of the agricultural revolution and the structural and dynamic similarities of agricultural groups no matter the species.

22. Edward O. Wilson, "One Giant Leap: How Insects Achieved Altruism and Colonial Life," *Bioscience* 58, no. 1 (2008): 17–25.

23. Wilson, "One Giant Leap."

Wilson and Hölldobler acknowledge that the story of the formation of insects into superorganisms is complex and the role of relatedness ambiguous. What is clear is that insect superorganisms have a lower "degree of within-colony relatedness" than is commonly assumed. This doesn't mean relatedness is entirely unimportant (there may be an element of kin selection at work), but it is certainly not the entire story. Relatedness in the *evolution* of eusociality is anything but clear; but what *is* clear is that relatedness is not a feature of eusociality once formed. There have been others who have also highlighted the problem of relatedness as it is used to explain insect colonies. Smith and Szathmáry, in *The Origins of Life*, point out that ants often have polygynous colonies where the average number of queens might be as high as a hundred, including queens not related. Ant foundress associations whereby unrelated queens cooperate to found new colonies are well documented. See John Maynard Smith and Eörs Szathmáry, *The Origins of Life: From Birth of Life to the Origins of Language* (Oxford: Oxford University Press, 2000). See also Lena Grinsted et al., "Subsocial Behaviour and Brood Adoption in Mixed-Species Colonies of Two Theridiid Spiders," *Naturwissenschaften* 99 (2012): 1021–30.

24. I am sympathetic to the effort to unseat the dominance of the selfish gene, and by extension kin selection and its extensions, and to reassert the importance of group selection. The use of the clarifying lens of altruism and its genetic connection have been important tools in this debate, but it is also possible that this orientation now prevents a more expansive exploration of group formation and its significance.

25. Martin Nowak et al., "The Evolution of Eusociality," *Nature* 466 (2010), 1060.

Wilson has stated (along with his coauthors) that among insect superorganisms "relatedness is better explained as a consequence rather than a cause of eusociality." Caste formation may occur through relatedness where sterile workers are essentially clones of the queen, but it is important to understand that this accommodates a division of labor in the production or reproduction of material life.

26. A few of the luminaries of entomology in this area are Bert Hölldobler and E. O. Wilson, Mark W. Moffett, Deborah M. Gordon, and Ana Duarte.

27. These are not the only genera of ants that practice agriculture, and there are many species of termites that do the same; but the leafcutter ants are clearly the most highly evolved.

28. Bert Hölldobler and Edward O. Wilson, *The Leafcutter Ants: Civilization by Instinct* (New York: W. W. Norton, 2011), 55.

29. Ulrich Mueller and Christian Rabeling, "A Breakthrough Innovation in Animal Evolution," *PNAS* 105, no. 14 (2008): 5287–88.

30. David Sloan Wilson, "Multilevel Selection and Major Transitions" in *Evolution the Extended Synthesis*, ed. Massimo Pigliucci and Gerd Müller (Cambridge, MA: MIT Press, 2010), 88.

31. Laurent Keller, *Levels of Selection in Evolution* (Princeton, NJ: Princeton University Press, 1999), 60.

32. Peter Corning, *Synergistic Selection: How Cooperation Has Shaped Evolution and the Rise of Humankind* (Hackensack, NJ: World Scientific, 2018), 91.

33. Peter Grimes. "Evolution and World-Systems: Complexity, Energy, and Form," *Journal of World-Systems Research* 23, no. 2 (2017), 690.

34. Wilson, *Social Conquest*, 183.

35. Hölldobler and Wilson map out in broad outline the necessary ingredients and unique sequence of events that they see in eusocial insect colonies: (1) formation of groups; (2) acquisition of preadaptations that form the ingredients that reinforce eusociality (a nest and defense of it); (3) the appearance of mutations that reinforce group persistence (this will manifest in very attenuated division of labor that might be accompanied with morphological differentiation); (4) emergent group traits that are reinforced through natural selection; and (5) multilevel selection that drives the evolution of superorganisms.

Chapter Two

1. Here the work of Carl Safina, Kevin Laland, Jane Goodall, Dian Fossey, and others come to mind.

2. Kevin Laland, "Evolution Unleashed: Is Evolutionary Science Due for a Major Overhaul—Or Is Talk of 'Revolution' Misguided?" *Aeon* (2018): 6.

3. Kevin Laland, *Darwin's Unfinished Symphony: How Culture Made the Human Mind* (Princeton, NJ: Princeton University Press, 2017), 174.

4. See Edward O. Wilson, *The Social Conquest of the Earth* (New York: Liveright, 2012) and Laland, *Darwin's Unfinished Symphony*.

5. See Peter Richerson and Robert Boyd, *Not by Genes Alone: How Culture Transformed Human Evolution* (Chicago: University of Chicago Press, 2005); Samuel Bowles and Herbert Gintis, *A Cooperative Species: Human Reciprocity and is Evolution* (Princeton, NJ: Princeton University Press, 2011); Laland, *Darwin's Unfinished Symphony*.

6. See Joseph Henrich, "Cultural Group Selection, Coevolutionary Processes and Large-scale Cooperation," *Journal of Economic Behavior and Organization* 53 (2004): 3–35; Richerson and Boyd, *Not by Genes Alone*; Bowles and Gintis, *A Cooperative Species*; Laland, *Darwin's Unfinished Symphony*.

7. In my work with John Gowdy, ultrasociality was used in the tradition of Donald Campbell, who defined it thus: "Ultrasociality refers to the social organization of a few species, including humans and some social insects, having a complex

division of labor, city-states, and an almost exclusive dependence on agriculture for subsistence." As previously noted, I now prefer to refer to those species that practice agriculture as economic superorganisms in deference to the power and integrity of the universal agricultural system. Humans and insects became ultrasocial with agriculture, but the system they are involved in is an economic superorganism. See Donald Campbell, "Legal and Primary-Group Social Controls," in *Law, Biology and Culture: The Evolution of Law*, ed. Margaret Gruter and Paul Bohannan (Berkeley: Bepress, 1982), 160.

8. The cooperation in insect superorganisms, including those that practice agriculture, is dissociated from humans by attributing insect cooperation to genetic relatedness. Having said that, it is good to be aware that there is also more complexity in the issue of relatedness (or not) in insect superorganisms. See Martin Nowak et al., "The Evolution of Eusociality," *Nature* 466 (2010): 1057–62.

9. Richerson and Boyd, *Not by Genes Alone*, 196.

10. See Herbert Gintis, *Individuality and Entanglement: The Moral and Material Bases of Social Life* (Princeton, NJ: Princeton University Press, 2017).

11. Mark Moffett, "Human Identity and the Evolution of Societies," *Human Nature* 24 (2013): 219–67.

12. See Christopher Boehm, *Moral Origins: The Evolution of Virtue, Altruism, and Shame* (New York: Basic Books, 2011); Bowles and Gintis, *A Cooperative Species*; Gintis, *Individuality and Entanglement*. To add to the complication of this conversation, it is entirely possible that the prosocial behaviors of humans evolve out of interaction with the more-than-human world and not solely out of the interaction with other humans. Take a trait such as empathy: an empathetic person is able to put him or herself in the shoes of another, thereby increasing their ability to be compassionate. There is no activity that requires putting yourself in the "shoes" of another more than hunting. One literally has to become the animal to know how to hunt it. Great compassion and connection emerge out of this complex human activity that may, in fact, spill over into the interaction between humans. The idea that we developed our unique human capacities by interacting with the more-than-human world must be reconsidered.

13. Laland, "Evolution Unleashed," 28. Hence for Laland, the evolution of the human mind is the result of a feedback between the refinement of copying with strategies such as "computation decision making, working, and long-term memory," which are associated with the "frontal and temporal lobes and prefrontal cortex." See also Laland, *Darwin's Unfinished Symphony*, 131.

14. Laland, *Darwin's Unfinished Symphony*, 28, 168.

15. Ana Duarte et al., "An Evolutionary Perspective on Self-Organized Division of Labor in Social Insects," *Annual Review of Ecology, Evolution, and Systematics* 42 (2011): 98.

16. David Wilson, "Multilevel Selection and Major Transitions" in *Evolution, the Extended Synthesis*, ed. Massimo Pigliucci and Gerd Müller (Cambridge, MA: MIT Press, 2010), 90.

17. I do understand the entropy law, but the dissipation of the sun's energy that grows plants is not something I am interested in.

18. John Gowdy and I referred to the division of labor as one of the economic drivers of agriculture in our coauthored work. It is probably more accurate to say that it was integral to an emergent agricultural system. See John Gowdy and Lisi Krall, "The Ultrasocial Origin of the Anthropocene," *Ecological Economics* 95 (2013): 137–47; John Gowdy and Lisi Krall, "Agriculture as a Major Evolution Transition to Human Ultrasociality," *Journal of Bioeconomics* 16 (2014): 179–202; John Gowdy and Lisi Krall, "The Economic Origins of Ultrasociality," *Behavioral and Brain Sciences* 39 (2016): e92; John Gowdy and Lisi Krall, "Disengaging from the Ultrasocial Economy," 39 (2016): e119; Lisi Krall, "The Economic Legacy of the Holocene," *Ecological Citizen* 2 (2018): 67–76; Lisi Krall, "Reckoning Our Ultrasocial Past," *Trumpeter* 31, no. 2 (2015): 102–111.

19. Laland, *Darwin's Unfinished Symphony*, 269.

20. See Deborah Gordon, "From Division of Labor to the Collective Behavior of Social Insects," *Behavioral Ecology and Sociobiology* 70 (2016): 1101–1108.

21. As previously stated, I have no doubt that the human attainment of culture was a major evolutionary transition for humans and that humans are uniquely intelligent. Nor do I have any doubt group selection for cultural capacity was a dominant force in the movement toward greater cooperation in humans. But cooperation is a complicated matter certainly in the realm of an economic system where structural scaffolding and feedback loops occur.

22. Laland, *Darwin's Unfinished Symphony*, 261, 269.

23. There are many others that stress culture as the necessary ingredient in the transition to agriculture. See Samuel Bowles and Jung-Kyoo Choi, "Coevolution of Farming and Private Property During the Early Holocene," *Proceedings of the National Academy of Sciences USA* 110 (22) (2013): 8830–35; Douglass North and Robert Thomas, "The First Economic Revolution," *Economic History Review* 30, no. 2 (1977): 229–41.

Bowles and Choi claim that property rights were a necessary institutional precondition for the transition to agriculture because without these rights farmers were not assured the benefit of their labor. In their analysis of property rights, a unique human arrangement enabled by culture is a decisive ingredient that ultimately helped move the winds of change in the direction of agriculture. Douglass North and Robert Thomas offer another institutional (cultural) variation on the theme of property rights and its role in the transition. Using the tools of marginal analysis, they ask what would lead a hunter to make the decision to cultivate instead of hunting and gathering. In their framework that decision at the margin is made on the basis of allocating the deployment of labor, presumably the scarce resource of preagricultural peoples, in a way that will maximize the return from it. North and Thomas similarly claim that the institution of property rights was essential in shifting things in the direction of agriculture because it guaranteed the return to labor input.

24. Claude Levi-Strauss tells us that humans had attained a very scientific mind through hunting and gathering long before science was born. Their ability to classify and identify species was expansive and exceeded the knowledge that was necessary to sustain them. And this knowledge, we are told, "lead[s] to scientifically valid results." Claude Levi-Strauss, *The Savage Mind* (Chicago: University of Chicago Press, 1962), 10.

25. James Scott, *Against the Grain: A Deep History of the Earliest States* (New Haven, CT: Yale University Press, 2017), 92.

26. Yuval N. Harari, *Sapiens: A Brief History of Humankind* (London: Vintage, 2011), 90–91. Wes Jackson has made the comment that if humans were meant to do agriculture they would have had longer arms.

27. Richerson and Boyd, *Not by Genes Alone*; Moffett, "Human Identity and the Evolution of Societies," 219; Laland, *Darwin's Unfinished Symphony*; Gintis, *Individuality and Entanglement.*

28. Genetic changes elicited by the force of the group might take time to show up and pressure for genetic change in humans is likely diminished by the presence of culture, institutions, and technology.

29. It is interesting that of the many extinctions now taking place, one of the most dramatic is the loss of the cultural and language diversity of the world.

30. Wilson, *The Social Conquest of the Earth*, 14–15. Wilson's perspective is that social insects achieved sociality "gradually, one innovation at a time" coevolving with the rest of the biosphere and becoming "vital elements of it." But for humans, evolution in the present *Homo sapiens* form took place very quickly. The human story is more dramatic in terms of time leaving humans without the ability to "coevolve with the rest of the biosphere." He would argue this is what makes humans an unecological species.

Chapter Three

1. See Clark Spencer Larsen, "The Agricultural Revolution as Environmental Catastrophe: Implications for Health and Lifestyles in the Holocene," *Quarterly International,* 150 (2006): 12–20; Yuval N. Harari, *Sapiens: A Brief History of Humankind* (London: Vintage, 2011); James Scott, *Against the Grain: A Deep History of the Earliest States* (New Haven, CT: Yale University Press, 2017).

2. Thomas S. Cox, "Crop Domestication and the First Plant Breeders," In *Plant Breeding and Farmer Participation,* ed. Salvatore Ceccarelli, Elcio Guimaraes, and Eva Weltzien (Rome: Food and Agricultural Organization, 2009), 1–26. It is important to note that the process of selection need not have been intentional. In the act of sowing a seed that had been harvested the previous season (as opposed to being picked off the ground) plants would have been selected inadvertently that did not have seeds that shattered. Shattering of seeds is important to wild plants

but selected against when harvesting. In this way domesticated seeds become wholly dependent on their relationship to humans.

3. Stands of wild grains harvested before agriculture (by the Natufians, for example) likely increased during the Holocene warming because climatic conditions were ideal for annual grains to flourish. Without the warmer and more stable climate conditions brought about by the Holocene it is unlikely that agriculture would have taken hold as a central strategy of physical life. In fact, it is likely that agriculture was experimented with, but never successful, before the Holocene. Another contributor to the agricultural revolution was the presence of soil carbon. Wes Jackson refers to soil carbon as one of the five pools of carbon that humans have accessed throughout history. Stored carbon soil added a boost to the success of agriculture because it enhanced the productivity of early agriculture. The store of soil carbon likely helped agriculture compete more fully with hunting and gathering at a time when the balance of material life could have stayed with the status quo. Apparently, soils at the beginning of the Holocene were very rich in carbon: estimates are that stored organic carbon was 33–60 percent lower in the late Pleistocene than in the Holocene. The added boost of a reserve of soil carbon was likely important given the many downsides to agriculture. See Joy McCorriston and Frank Hole, "The Ecology of Seasonal Stress and the Origins of Agriculture in the Near East," *American Anthropologist, New Series* 93, no. 1 (1991): 46–69; Peter Richerson, Robert Boyd, and Robert Bettinger, "Was Agriculture Impossible During the Pleistocene but Mandatory During the Holocene? A Climate Change Hypothesis," *American Antiquity* 66 (2001): 387–411; Ofer Bar-Yosef, "The Natufian Culture in the Levant, Threshold to the Origins of Agriculture," *Evolutionary Anthropology* 6 (1998): 159–77; Wes Jackson "Five Carbon Pools," in *The Energy Reader: Overdevelopment and the Delusion of Endless Growth*, ed. Tom Butler, Daniel Lerch, and George Wuerthner (Sausalito, CA: Foundation for Deep Ecology, 2012), 27–32; David Beerling, "New Estimates of Carbon Transfer to Terrestrial Ecosystems Between the Last Glacial Maximum and the Holocene," *Terra Nova* 11 (1999): 162–67.

4. I am not claiming that agriculture did not have any effect on genetic selection. Lactose tolerance in adults is one of the most obvious changes that has occurred since the onset of agriculture.

5. According to Samir Okasha, "Emergent characters are often complex, adaptive features of collectives, which it is hard to imagine evolving except by selection at the collective level." See Okasha, *Evolution and the Levels of Selection*, 112.

Clearly there are complex dialectical processes at work. It should also be noted that role differentiation by genes is only one aspect of the determinant of division of labor in insects. Apparently, in some cases, they are able to move from one task to another as needed. Deborah Gordon, "From Division of Labor to the Collective Behavior of Social Insects," *Behavioral Ecology and Sociobiology* 70 (2016): 1101–1108.

6. The literature on division of labor in social insects is extensive. I cite some examples here, but this is not an exhaustive list of the work done on this

topic. In addition to the work done by Höolldobler and Wilson, some of the other specialists in organization of superorganisms are Samuel Beshers and Jennifer Fewell, "Models of Division of Labor in Social Insects," *Annual Review of Entomology* 46 (2001): 413–40; Fewell, "Social Insect Networks," *Science*, 301 (2003): 1867–70; Ted Schultz and Sean Brady, "Major Evolutionary Transition in Ant Agriculture," *Proceedings of the National Academy of Sciences* 105 (2008): 5435–40; Duarte at al., "An Evolutionary Perspective on Self-Organized Division of Labor in Social Insects," 91–110; C. Tate Holbrook, Phillip Barden, and Jennifer Fewell, "Division of Labor Increases with Colony Size in Harvester Ant *Pogonmyrmex californicus*," *Behvioral Ecology* 22, Issue 5 (2011): 960–66; Deborah Gordon "From Division of Labor to Collective Behavior; Robert Jeanne, "Division of Labor is Not a Process or a Misleading Concept," *Behavioral Ecology and Sociobiology* 70, no. 7 (2016): 1109–12; Jacobus Boomsma and Nigel Franks, "Social Insects: From Selfish Genes to Self-Organisation and Beyond," *Trends in Ecology and Evolution* 21, no. 6 (2006): 303–308; Simon Robson and James Traniello, "Division of Labor in Complex Societies: A New Age of Conceptual Expansion and Integrative Analysis," *Behavioral Ecology and Sociobiology* 70 (2016): 995–98. There is some disagreement among sociobiologists who study insects over terminology and mechanisms of engagement of the division of labor. For example, Deborah Gordon argues that "Division of labor is a misleading way to describe the organization of tasks in social insect colonies, because there is little evidence for persistent individual specialization in task . . . and tends to focus attention on differences among individuals in internal attributes." Gordon claims the use of the division of labor "distracts from . . . an understanding of how individuals interact with each other and their environments." Gordon, "From Division of Labor to Collective Behavior," 1101. I acknowledge that Gordon identifies an important distinction—the coordination of collective work in different environments is a somewhat different question than asking why and how individuals come to be relegated to certain tasks or roles. In this document I use the division of labor to talk about both, but the emphasis, as with Gordon, is on the fact of it and the conditions that extend it. Why certain individuals end up doing certain tasks is a different question than the question of the existence of tasks, their connection to one another in common purpose, and the conditions under which they arise.

7. Martin Nowak et al., "The Evolution of Eusociality," *Nature* 466 (2010): 1057–62, The authors tell us: "The division of labor appears to be the result of a pre-existing behavioral ground plan, in which solitary individuals tend to move from one task to another only after the first is completed . . . the algorithm is readily transferred to the avoidance of a job already being filled by another colony member." The division of labor is seemingly "automatic" and relies on plasticity or totipotency and/or small differences in "predisposition and experience."

See also Bert Hölldobler and Edward O. Wilson, *The Superorganism* (New York: W. W. Norton, 2009), 87.

Hölldobler and Wilson are very clear about the division of labor in ant superorganisms: "During the evolution of insect societies, from over 100 million years ago forward, division of labor appears to have arisen very easily. In the case of ant species, it is all but automatic . . . the smallest differences in predisposition and experience can be amplified into a non-reproductive division of labor . . . It is notable that the different roles of the reproducing parents and their non-reproductive offspring are not genetically determined."

8. Hölldobler and Wilson, *The Superorganism*, 91.

9. It would appear that the more reliable the food source (as in the case of the leafcutter ants who are feeding their fungal gardens with fresh vegetation) the greater the morphological differentiation or the greater the hardening of the division of labor.

10. Mark Moffett, *Adventures Among Ants: A Global Safari with a Cast of Trillions* (Berkeley: University of California Press, 2010), 172.

11. Simon Robson and James Traniello, "Division of Labor in Complex Societies: A New Age of Conceptual Expansion and Integrative Analysis," *Behavioral Ecology and Sociobiology* 70 (2016): 995–98.

12. Tim Flannery, "The Superior Civilization," *New York Review of Books*, February 26, 2009.

13. See John Gowdy and Lisi Krall, "The Ultrasocial Origin of the Anthropocene," *Ecological Economics* 95 (2013): 137–147; John Gowdy and Lisi Krall, "Agriculture as a Major Evolution," *Journal of Bioeconomics* 16 (2014): 179–202; John Gowdy and Lisi Krall, "The Economic Origins of Ultrasociality," *Behavioral and Brain Sciences* 39 (2016), e92.

In our published work we claimed that the division of labor was one of the powerful "economic drivers" in the formation of agriculture. It is a powerful driver of system formation but in the end the emphasis must be on the system created.

14. Bruce Wexler, *Brain and Culture* (Cambridge, MA: MIT Press, 2006), 32.

As Wexler tells us: "High levels of plasticity in the relationship between structure and function persist for years in the structures that most distinguish the human brain from those of other primates. This creates an unprecedented opportunity for environmental shaping of uniquely human aspects of brain function."

15. Understand that the efficiencies inherent in the division of labor may be counteracted in some sense by social norms. Such is the gender division of labor where individuals are allocated to a selection of jobs based on gender and regardless of their interest and capability.

16. Adam Smith, *An Inquiry into the Nature and Causes of the Wealth of Nations* (Chicago: University of Chicago Press, 1976), 7. Smith reveals reservations about the detailed division of labor where he feared that doing the same mindless tasks over and over again could make humans "as ignorant and stupid as it was possible for them to be" (302–303).

17. See also Joseph Schumpeter, *History of Economic Analysis* (New York: Oxford University Press, 1986), 56. Schumpeter characterizes Plato's reference to the division of labor by saying, "He elaborates on this eternal commonplace of economics with unusual care."

18. I cannot resolve here the full complexity of the universal aspect of the division of labor. It is clear that a capacity for differentiation appears to be quite universal in the biological world—and is something that certainly extends beyond culture. The evolutionary biologist David Sloan Wilson tells us, "The fact that cells of multicellular organisms are totipotent at the genetic level but also capable of extreme specialization through the differential expression of genes, provides a useful frame of reference for thinking about human division of labor." See David S. Wilson "Laying the Foundation for Evonomics," Open Peer Commentary to Gowdy and Krall in "The Economic Origins of Ultrasociality," *Behavioral and Brain Sciences* 39 (2016): 40. Mary Midgley tells us that specialization is one of the "intelligible principles" of collective species life. See Mary Midgley, *Beast and Man: The Roots of Human Nature* (Abingdon, UK: Routledge, 2002), 18.

19. M. W. Moffett, *Adventures Among Ants: A Global Safari with a Cast of Trillions* (Berkeley: University of California Press, 2010), 174–75.

20. It is important to understand that in the realm of evolution something small may end up more significant. Especially in the context of the development of a system there are synergies and dialectics that create significant feedback loops that must be recognized.

21. Smith, *Wealth of Nations*, 13. One thing Smith got wrong in his analysis was his notion that markets were universal and there was an innate tendency in human to "truck, barter, and exchange."

22. Karl Marx, "The German Ideology: Part I," in *The Marx-Engels Reader*, ed. Robert C. Tucker (New York: W. W. Norton, 1972), 114, 121, 124.

23. Later economists in the Marxian tradition have explored in detail the division of labor under industrial capitalism. See Harry Braverman, *Labor and Monopoly Capital* (New York, Monthly Review Press, 1974); and Richard Edward, *Contested Terrain: The Transformation of the Workplace in the Twentieth Century* (New York: Basic Books, 1979).

24. Marx simply considered agriculture an "undeveloped stage of production" where "the division of labor is still very elementary and is confined to a further extension of the natural division of labor existing in the family." Marx, "The German Ideology," 115.

25. Smith, *Wealth of Nations*, 17.

26. Emile Durkheim, *The Division of Labor in Society* (New York: Free Press, 1964), 40–44. Other social scientists have similarly explored the origins of this propensity. For example, Karl Marx attributed the division of labor to a "physiological foundation" and finds its origins in the family: "Within a family, and after further development with a tribe, there springs up naturally a division of labour, caused

by differences of sex and age, a division that is consequently based on a purely physiological foundation." Karl Marx, *Capital: A Critique of Political Economy, Vol. I: The Process of Capitalist Production* (New York: International Publishers, 1979), 351. George Herbert Mead recognizes the similarities between human societies and the societies of social insects. Mead tells us: "The insects reveal a very curious development. We are tempted to be anthropomorphic in our accounts of the life of bees and ants, since it seems comparatively easy to trace the organization of human community in their organizations. There are different types of individuals with corresponding functions, and a life-process as analogous to a human society. We have not, however, any basis as yet for carrying out the analogy in this fashion because we are unable to identify any system of communication in insect societies." Had Mead had all of the information available today about the formation and structure of insect superorganism his analysis might have been inclined differently: his conclusions are not substantiated by the fact that there are forms of communication between individual insects in a colony. George Herbert Mead, *Mind, Self and Society: The Definitive Edition* (Chicago: University of Chicago Press, 2015), 230–31.

27. Peter Corning, *Synergistic Selection: How Cooperation Has Shaped Evolution and the Rise of Humankind* (Hackensack, NJ: World Scientific, 2018), 50.

28. Price and Bar-Yosef comment on the lack of consensus about the "cause" of the momentous change to agriculture: "There is as yet no single accepted theory for the origins of agriculture, rather, there is a series of ideas and suggestions that do not quite resolve the questions." T. Douglas Price and Ofer Bar-Yosef, "The Origins of Agriculture: New Data, New Ideas, An Introduction to Supplement 4," *Current Anthropology* 52, Supplement (October 2011), 168. Some anthropologists have pushed the theory that population pressure drove the agricultural revolution partly because the agricultural revolution occurred independently in different parts of the world; so, in the search for universal drivers, population pressure was a logical candidate. See Lewis R. Binford, "Post-Pleistocene Adaptations," in *New Perspectives in Archaeology*, ed. Lewis Binford and S. R. Binford (Chicago: Aldine, 1968), 313–41. Cohen tells us that population pressure meant that people were simply required to alter their food strategy. See Mark Cohen, *The Food Crisis in Prehistory: Overpopulation and the Origins of Agriculture* (New Haven, CT: Yale University Press, 1977). Population changes in the distant past are clearly difficult to demonstrate as Price and Bar-Yosef point out in *Overpopulation and the Origins of Agriculture* there is little evidence for population pressures in the place where agriculture first appeared. The population factor is even more confusing because it is clear that for many reasons sedentary life raises fertility rates. It is possible that any partial switch to grain cultivation or greater dependency on the harvesting of wild grains could have contributed to fertility changes even short of permanent settlements if people simply stayed put for longer periods. I clearly prefer to think of agriculture from the perspective of an emergent and universal system and then to concentrate on the structure, dynamic, and power of that system.

Chapter Four

1. Let me reiterate the universal processes. Engaging an agricultural system is not simply an individual matter: it is also a collective one that resides in the capacity of a species to collectively configure themselves around cultivation. Many species have similar collective capabilities. The resulting system is one of increased population, increased division of labor, increased output in an upward and self-referential spiral.

2. Civilization may be simply the veneer of a downward spiral into the economic superorganism where a duality has been created between humans and the more-than-human world around a self-referential expansionary human system that now threatens both the collapse of the human economic superorganism and the temporary annihilation of the more-than-human world. Temporary because in the long arc of Earth's history, the Earth will recover. This will be taken up more systematically in the next chapter.

3. For discussions of the various problems with agriculture, see Mark Cohen and Jillian Crane-Kramer, *Ancient Health: Skeletal Indicators of Agricultural and Economic Intensification* (Gainesville: University of Florida Press, 2007); Clark Spencer Larsen, "The Agricultural Revolution as Environmental Catastrophe: Implications for Health and Lifestyles in the Holocene," *Quarterly International* 150 (2006): 12–20; Patricia Lambert, "Health versus Fitness," *Current Anthropology* 50, no. 5 (2009): 603–608; Yuval N. Harari, *Sapiens: A Brief History of Humankind* (London: Vintage, 2011); James Scott, *Against the Grain: A Deep History of the Earliest States* (New Haven, CT: Yale University Press, 2017); John Gowdy and Lisi Krall, "Agriculture as a Major Evolution Transition to Human Ultrasociality," *Journal of Bioeconomics* 16 (2014): 179–202.

4. Jared Diamond, "The Worst Mistake in the History of the Human Race," *Discover*, May 1987: 64–66.

5. Harari, *Sapiens*, 89.

6. Paul Sears, *Deserts on the March* (Norman: University of Oklahoma Press, 1959), 9.

7. John Gowdy and Lisi Krall, "The Ultrasocial Origin of the Anthropocene," *Ecological Economics* 95 (2013): 137–47; Gowdy and Krall, "Agriculture as a Major Evolution Transition"; John Gowdy and Lisi Krall, "The Economic Origins of Ultrasociality," *Behavioral and Brain Sciences* 39 (2016), 1–60; Lisi Krall, "The Economic Legacy of the Holocene," *Ecological Citizen* 2 (2018), 67–76; Lisi Krall, "Reckoning Our Ultrasocial Past," *Trumpeter* 31, no. 2 (2015): 102–111.

8. Harari, *Sapiens*.

9. Robert Heilbronner, *The Worldly Philosophers* (New York: Simon and Schuster, 1999), 18.

10. Steven Pinker, *Enlightenment Now: The Case for Reason, Science, Humanism, and Progress* (New York: Penguin, 2018), 23.

11. If one looks at the population dynamic after agriculture one sees that the rate of population growth increased after agriculture began. But for the first five thousand years the rate of growth appears to be lower than it became subsequently. James Scott tells us that this was due to the dramatic rise in mortality caused by the crowding diseases associated with sedentary life with domesticated animals that accompanied agriculture. The birth rates were high but so were the death rates. See Scott, *Against the Grain*.

12. Peter Richerson, Robert Boyd, and Robert Bettinger, "Was Agriculture Impossible During the Pleistocene but Mandatory During the Holocene? A Climate Change Hypothesis," *American Antiquity* 66 (2001): 387–411.

13. Wes Jackson, "Five Carbon Pools," in *The Energy Reader: Overdevelopment and the Delusion of Endless Growth*, ed. Tom Butler, Daniel, Lerch, and George Wuerthner (Sausalito, CA: Foundation for Deep Ecology 2012), 27–32.

14. Scott, *Against the Grain*.

15. Harari, *Sapiens*, 115.

Harari recognizes the characteristics of grains when he tells us in dramatic fashion that grains made humans, but of course the truth is that it was a coevolutionary process. And he actually gives a lot of power in the process to human imagination. Yuval Harari retreats to human imagination to explain the rise of state societies. In his words: "While human evolution was crawling at its usual snail's pace, the human imagination was building astounding networks of mass cooperation, unlike any other ever seen on earth." Just to remind the reader—the mass cooperation and civilizations attained by humans with agriculture had already been attained by a multitude of ant and termite species many millions of years before our ancestors descended from the trees. Perhaps Harari gives more to human imagination than is owed.

16. Wes Jackson, *Consulting the Genius of the Place* (Berkeley, CA: Counterpoint, 2010), 87.

17. This analysis does not concentrate on the animal husbandry, which accompanied agriculture and was not insignificant. Animal husbandry was essential for nutrient recycling, but domesticated animals also increased the likelihood of disease. And pastoralism was always an option for upland soil that could not easily be irrigated or was otherwise (due to innate fertility) not good for cultivation. Pastoralism increased overgrazing, adding to the problem of siltation of irrigations systems. Paul Sears tells us, "The history of early civilization can be written largely in terms of these two great inventions in living—the pastoral life of the dry interior and settled agricultural of the well-watered regions." Paul Sears, *Deserts on the March* (Norman: University of Oklahoma Press, 1959), 9.

18. Gowdy and Krall, "The Ultrasocial Origin"; Gowdy and Krall, "Agriculture as a Major Evolution"; Gowdy and Krall, "The Economic Origins of Ultrasociality."

19. Again it is important to appreciate the efficiencies garnered through a division of labor elaborated in the last chapter. This is something emphasized and

noted by both economists and entomologists as well as evolutionary biologists. Adam Smith's famous tome on capitalism, *The Wealth of Nations*, begins with three chapters on the division of labor. And E. O. Wilson and Bert Hölldobler's *The Superorganism* allocated an entire section to the division of labor. The benefits of a division of labor are widely acknowledged across disciplines. See Bert Hölldobler and Edward O. Wilson, *The Superorganism* (New York: W. W. Norton, 2009).

20. Scott, *Against the Grain*, 92.

Harari makes a similar claim when he says that "the Agricultural Revolution was history's biggest fraud." Harari, *Sapiens*, 90. He then thinks about "the Agricultural Revolution" from the viewpoint of wheat and in doing so talks about the way wheat manipulated "*Homo sapiens* to its advantage." Harari takes the coevolutionary dynamic to an extreme and tells us that wheat became one of "the most successful plants in the history of the earth . . . by manipulating *Homo sapiens* to its advantage." He continues: "Within a couple of millennia, humans in many parts of the world were doing little from dawn to dusk other than taking care of plants. . . . Wheat didn't like sharing its space, water and nutrients with other plants, so men and women labored long days weeding under the scorching sun. Wheat got sick, so Sapiens had to keep a watch out for worms and blight. Wheat was attacked by rabbits and locust swarms, so the farmers built fences and stood guard over the fields. Wheat was thirsty, so humans dug irrigation canals or lugged heavy buckets from the well to water it. . . . The body of *Homo sapiens* had not evolved for such tasks." Harari is a bit dramatic in his characterization because it was at least a coevolutionary process, but his point is well taken.

21. Wes Jackson, *Nature as Measure* (Berkeley, CA: Counterpoint, 2011).

22. As the anthropologist Sharon Steadman reminded me, as family size grows, "larger families can employ the older children in many tasks associated with agriculture. . . . As well, multigenerational family structure either develops or is reinforced so that elders can contribute by watching the little kids while mom/older children/dad do work away from home." Thus is it possible to see that complexity even emerges around family structure, which comports with the more expansive complexity emerging in the whole of society. Personal conversation with Sharon Steadman, January 2020.

23. Scott, *Against the Grain*, 97.

Scott tells us that "epidemiologically, this was perhaps the most lethal period in human history." He refers to the sedentary places of habitation ("multispecies resettlement camps") and claims they were lethal.

24. There is some indication that at the inception of agriculture there may have been an orientation toward female-based religious symbolism; however, this was temporary, giving way to patriarchy.

25. Joseph Tainter, "Sustainability of Complex Societies," *Futures* 27, no. 4, 397–407.

26. It is interesting to note that the similarities in the civilizations that arose out of agriculture are striking. Ronald Wright tells us of the discovery of the Americas in the 1500s: "Two cultural experiments, running in isolation for 15,000 years or more, at last came face to face. Amazingly, after all that time, each could recognize the other's institutions. When Cortés landed in Mexico he found roads, canals, cities, palaces, schools, law courts, markets, irrigation works, kings, priests, temples, peasants, artisans, armies, astronomers, merchants, sports, theatre, art, music, and books. High civilization, differing in detail but alike in essentials, had evolved independently on both sides of the earth." Ronald Wright, A *Short History of Progress* (Cambridge, MA: DaCapo, 2004), 50.

Chapter Five

1. Bill McKibben, *Falter* (New York: Henry Holt, 2019). McKibben frets about the possibilities of new genetic engineering technology eventually having the capability of directing human evolution. He rightly fears this dystopic future but never considers how fundamentally altered humans can become without altering a single strand of their DNA.

2. This again is Wes Jackson's phraseology in "Five Carbon Pools." See Wes Jackson "Five Carbon Pools," in *The Energy Reader: Overdevelopment and the Delusion of Endless Growth*, ed. Tom Butler, Daniel Lerch, and George Wuerthner (Sausalito, CA: Foundation for Deep Ecology 2012), 27–32.

3. Paul Shepard, *Nature and Madness* (San Francisco: Sierra Club, 1982).

4. Wendell Berry, "The Peace of Wild Things," in *The Peace of Wild Things: And Other Poems* (London: Penguin, 2018).

5. There is consensus that humans evolved culture as a mechanism of quick adaptation during the dramatic climatic changes of the Pleistocene. See, for example, Peter Richerson and Robert Boyd, *Not by Genes Alone: How Culture Transformed Human Evolution* (Chicago: University of Chicago Press, 2005); Kevin Laland, *Darwin's Unfinished Symphony: How Culture Made the Human Mind* (Princeton, NJ: Princeton University Press, 2017); Samuel Bowles and Herbert Gintis, *A Cooperative Species: Human Reciprocity and its Evolution* (Princeton, NJ: Princeton University Press, 2011).

6. Stephen J. Gould, *Ever Since Darwin* (New York: W. W. Norton, 1977), 251.

7. Paul Shepard, *Nature and Madness* (San Francisco: Sierra Club, 1982). See also Paul Shepard, *A Paul Shepard Reader: The Only World We've Got*, ed. Paul Shepard (San Francisco: Sierra Club. 1996); Claude Levi-Strauss, *The Savage Mind* (Chicago: University of Chicago Press, 1970).

8. That is not to say there wasn't significant ecological destruction before the exponential growth curves of the fossil fuel era began their upward reach. The collapse

of previous agricultural civilizations has been well documented—in Mesopotamia, large civilizations that grew up surrounding the Mediterranean, the Mayans, the Incas, the Anasazi, those in the Yellow River—for starters. Many of the landscapes of the Near East in particular have never recovered from annual grain agriculture. See Paul Sears, *Deserts on the March* (Norman: University of Oklahoma Press, 1959). Despite this ecological decline one cannot argue that the wild was eradicated from the Earth.

9. Stanley Diamond, *In Search of the Primitive: A Critique of Civilization* (New Brunswick, NJ: Transaction, 1974), 119.

10. As a reminder, as anatomically modern humans we practiced the hunting and gathering mode of production for two hundred thousand to three hundred thousand years before agriculture took hold.

11. Marshall Sahlins, *Stone Age Economics* (London: Routledge, 1974).

12. The preconception that the hunter-gatherer worked from dawn until dusk to eke out a minimal existence is so ubiquitous that it has all but become accepted knowledge. By all accounts hunters and gatherers roamed and gathered for about six hours per day to obtain what they needed and took the rest of the day in what we would consider leisure. This is not exactly the schedule of people teetering on the brink of starvation. With the preconception of scarcity the narrative that human intelligence allowed us to overcome this horrible and relentless specter is elevated. I think erroneous interpretations of human prehistory suffer from the lens of the present extrapolated to the past.

13. EROEI (energy return on energy invested). See Charles A. S. Hall, "Energy Return on Investment," in *The Energy Reader: Overdevelopment and the Delusion of Endless Growth,* ed. Tom Butler, Dan Lerch, and George Wuerthner (Sausalito, CA: Watershed Media, 2012), 62–68.

14. Marshall Sahlins, "The Original Affluent Society," in *Limited Wants Unlimited Means: A Reader on Hunter-Gatherer Economics and the Environment*, ed. John Gowdy (Washington, DC: Island, 1998), 5–6.

15. See Diamond, *In Search of the Primitive.*

16. Samuel Bowles, "Warriors, Levelers, and the Role of Conflict in Human Social Evolution," *Science* 336 (2012): 876–79.

17. Richard Lee, "Non-Capitalist Work: Baseline for an Anthropology of Work or Romantic Delusion?" *Anthropology of Work Review* 18, no. 4 (1998): 9–13.

18. James Scott, *Against the Grain: A Deep History of the Earliest States* (New Haven, CT: Yale University Press, 2017).

19. Kirkpatrick Sale, *After Eden: The Evolution of Human Domination* (Durham, NC: Duke University Press, 2006), 29.

20. Eileen Crist, *Abundant Earth* (Chicago: University of Chicago Press, 2019).

21. This is not an accepted explanation in physical anthropology and archeology circles, but it is a narrative frequently cited nonetheless.

22. Meat was important to human evolution especially for brain development, which is a high-energy evolutionary proposition. Before the advent of hunting, scavenging was the order of the day as far as meat provisioning was concerned, and

observation was the key to scavenging likely in following the circling of buzzards around dead animals. I have wondered whether the migration out of Africa may have been the result of following the annual migration paths of buzzards, which lead humans to carcasses. Access to energy in the form of cooking along with meat helped to increase the cranial capacity of humans over the long arc of human evolution.

23. Gerda Lerner, *The Creation of Patriarchy* (New York: Oxford University Press, 1986), 17. In many hunter-gatherer cultures, 80 percent of diet by weight is plants. See Lee, "Non-Capitalist Work."

24. There is a great deal of controversy about the Pleistocene overkill, and there are numerous scholars that say the Pleistocene overkill is overemphasized as a cause of the extinction of big mammals that occurred twelve thousand years ago, especially given the contribution of the dramatic climatic changes that were occurring.

25. See Scott, *Against the Grain*. He discusses the spread of the "crowding diseases" after the onset of agriculture: tuberculosis, mumps, measles, influenza, typhus, and cholera that were a formidable check on population growth. Population growth would have been much higher with the advent of agriculture without the presence of these diseases.

26. Ronald Wright tells us that "by 3,000 years ago, civilization had arisen in at least seven places: Mesopotamia, Egypt, the Mediterranean, India, China, Mexico, and Peru. Archaeology shows that only about half of these had received their crops and cultural stimuli from others. The rest had built themselves up from scratch without suspecting that anyone else in the world was doing the same." Ronald Wright, A *Short History of Progress* (Cambridge, MA: DaCapo, 2004), 65.

27. Again see Joseph Tainter, "Sustainability of Complex Societies." *Futures* 27, no. 4 (1995): 397–407.

28. Scott, *Against the Grain*, 90–91.

29. Bill Vitek, "Dandelions are Divine," *Ecological Citizen* 2 (2019): 191.

30. Bill Vitek, "God or Nature: Desire and the Quest for Unity," *Minding Nature* 4, no. 2 (2011): 21.

31. I offer an unapologetically materialist perspective because I am saying without equivocation that it was the altered material system that altered the worldview. That altered worldview came to be embodied in Enlightenment thinking where it is believed that humans can solve any problem they confront if they just set their mind to it. This is the thread in Steven Pinker's book *Enlightenment Now* where he explicitly chastises anyone who would question the wisdom of the perspective that humans are on the road to progress. Pinker labels those who question progress as the "chattering class." I am a member.

Chapter Six

1. Simon Lewis and Mark Maslin, *The Human Planet: How We Created the Anthropocene* (New Haven, CT: Yale University Press, 2018), 149.

2. The steam engine was perfected by James Watt in 1769—a mere seven years before Adam Smith published his five-hundred-page tome on capitalism.

3. Thorstein Veblen, *Absentee Ownership and the Business Enterprise in Recent Times* (New York: Sentry, 1964), 37.

4. See Jason Moore, "The Rise of Cheap Nature" in *Anthropocene or Capitalocene? Nature, History, and the Crisis of Capitalism,* ed. Jason Moore (Oakland, CA: PM, 2016), 78–115. Moore is part of the World Systems tradition that comes out of sociology. His work revolves around a universal dictum: humans have changed the Earth, and the Earth has changed humans; he approaches capitalism as a world ecology. While I recognize this dialectic, the balance of that dialectic changed dramatically with the agricultural revolution, which is something Moore doesn't recognize. I will expand on Moore's thinking in the next chapter.

5. Andreas Malm, *Fossil Capital: The Rise of Steam Power and the Roots of Global Warming* (London: Verso, 2016).

6. See for example, E. K. Hunt, *Property and Prophets: The Evolution of economic Institutions and Ideologies* (Armonk, NY, and London: M. E. Sharpe, 2003).

7. Robert Heilbronner, *The Worldly Philosophers* (New York: Simon and Schuster, 1999), 3.

8. World Systems literature concentrates on the emergence of capitalism long before the sixteenth and seventeenth centuries. For further reading, see Janet Abu-Lughod, "The Shape of the World System in the Thirteenth Century," *Studies in Comparative International Development* 22, no. 4 (1988): 3–24; Andre Frank, "A Theoretical Introduction to 5000 Years of World System History," *Review* 13 (1990): 155–248; Barry Gills and Andre Frank, "5000 Years of World System History: The Cumulation of Accumulation," in *Core/Periphery Relations in Pre-capitalist Worlds,* ed. Christopher Chase-Dunn and Thomas D. Hall (Boulder, CO: Westview, 1991), 67–112; Giovanni Arrighi, *Adam Smith in Beijing: Lineages of the Twenty-First Century* (London: Verso, 2007).

I appreciate both the World Systems tradition as well as the tradition encapsulated by the work of Eric Hobsbawm, Edward P. Thompson, and Ellen Meiksins Wood, among others. They recognize the ingredients necessary to capitalism as it fully took form in Europe and matured in England in the eighteenth century. See Eric Hobsbawm, *The Age of Revolution 1789–1848* (London: Weidenfeld and Nicolson, 1962); Edward P. Thompson, *The Making of the English Working Class* (New York: Vintage, 1963); Ellen Meiksins Wood, *The Origin of Capitalism* (New York: Monthly Review, 1999).

9. Karl Polanyi, *The Great Transformation* (Boston: Beacon, 1957), 68.

10. It is essential to understand that capitalism is chiseled into contemporary form out of feudal Europe, especially in Great Britain. This analysis does not engage the controversies found in General Systems Theory about why capitalism did not arise in Asia. For further reading on that topic see Giovanni Arrighi, *Adam Smith in Beijing: Lineages of the Twenty-First Century* (London: Verso, 2007).

11. To give an example of how ill-equipped the manorial world was for trade I quote Heilbronner who describes a merchant in Germany in 1550: "Andreas Ryff, a merchant . . . is troubled by the nuisances of the times: As he travels he is stopped approximately once every 10 miles to pay a customs toll; between Basel and Cologne he pays 31 levies. And that is not all. Each community he visits has its own money, its own rules and regulations, its own law and order. In the area around Baden alone there are 112 different measures of length, 92 different square measure, 65 different dry measures, 163 measures for cereals, and 123 for liquids, 63 special measures for liquor, and 80 different pound measures." Heilbronner, *Worldly Philosophers*, 22.

12. For a lengthier but accessible understanding of the complexity of the transition and the many elements involved, see Emery K. Hunt, *Property and Prophets: The Evolution of Economic Institutions and Ideologies*, 7th ed. (Armonk, NY: M. E. Sharpe, 2003).

13. Please understand that trade had been increasing since the beginning of agriculture and certainly with the rise of state societies.

14. Heilbronner, *Worldly Philosophers*, 25.

15. See Jason Moore, "Ecology, Capital, and the Nature of Our Times: Accumulation and Crisis in the Capitalist World-Ecology," *American Sociology Association* 18, no. 1 (2011): 108–147.

16. Lewis and Maslin, *Human Planet*, 150.

17. Alfred W. Crosby Jr., *The Colombian Exchange* (Westport, CT: Greenwood, 1972).

18. Lewis and Maslin, *Human Planet*, 191. In their book they have a wonderful map that shows the expanding trade routes emerging almost entirely by sea; they call this a New Pangea, 162–63.

19. Eric Hobsbawm, *The Age of Revolution: Europe 1789–1848* (London: Weidenfeld and Nicolson, 1962), 149.

20. Hobsbawm, *Age of Revolution*, 153.

21. It is important to understand that land relationships varied by region and country.

22. Hobsbawm, *Age of Revolution*, 34.

23. Andreas Malm, *Fossil Capital: The Rise of Steam Power and the Roots of Global Warming* (London: Verso, 2016), 45.

24. Wood, *The Origin of Capitalism,*

25. Of course, profits are always realized in the act of selling, but the foundation for that realization takes place in production.

26. The same would be the case in iron making, sugar refining, pottery, bricks, glass, and mining, and many other lines of production.

27. This form of securing labor abounded on sugar plantations as well. Here processing took place in colonial regions.

28. Thompson, *Making of the English Working Class*, 198.

29. In Marxian terminology this is called surplus value.

30. Marx is quoted in Hunt, *Property and Prophets*, 111.

31. According to Andreas Malm, fossil fuel did augment the energy use in households for heating and cooking but when industrial production took off in textiles it was not fueled by fossil fuel.

32. Polanyi, *The Great Transformation*, 163, 178.

33. Jason W. Moore discusses capitalism's world ecology. It is encapsulated in his discussion of cheap natures.

34. Malm, *Fossil Capital*, 131.

35. Malm, *Fossil Capital*, 136.

36. Lynn White Jr., *Medieval Technology and Social Change* (Oxford: Oxford University Press, 1964), 39.

37. Malm, *Fossil Capital*, 145.

38. Malm, *Fossil Capital*, 151.

39. Harry Braverman (extending the work of Marx) explored the division of labor specific to the advanced form of industrial capitalism—monopoly capital. In his classic work *Labor and Monopoly Capital* Braverman clearly understands the proclivity of capitalism toward the efficiencies and control of the labor process, and he understands very clearly that this tendency reaches its apogee in industrial capitalism fueled with fossil fuel. Workers became increasingly interdependent not only around a social division of labor but now in a highly mechanized detailed division of labor. Harry Braverman, *Labor and Monopoly Capital* (New York: Monthly Review, 1974).

40. Thompson, *The Making of the English Working Class*, 243. It is clear that the present fear of technology replacing people is not new.

41. The truth, of course, is that we had been riding exponential growth since the advent of agriculture, but growth rates were relatively modest until the industrial revolution. This is evident in population growth that goes from about six million at the beginning of the agricultural revolution to one billion by the beginning of the nineteenth century to the present 7.8 billion, a little more than two hundred years later.

42. David Landes, *The Unbound Prometheus: Technological Change and Industrial Development in Western Europe from 1750 to the Present* (Cambridge: Cambridge University Press, 1969), 97.

43. I fully recognize that all life needs energy and that access to energy might provide an evolutionary advantage, but there are different energy systems.

44. This is a simplified interpretation of a Marxian perspective. This is but one rendition of the contradictions of the system. Keynes, too, identified contradictions. His analysis concentrated on the inequality that might emerge between savings and investment that became especially formidable as capitalism matured.

45. Perhaps the most well known of the economists to understand and explore the nature of class conflict in capitalism was Karl Marx. Marx clearly understood the purpose of production under capitalism, which is to make a profit and to have

the right to it by virtue of private property. In this he describes the particular social relations underlying capitalism. Marx was aware that this purpose and the tensions embodied therein will take hold and reveal themselves through class conflict and deepening depression.

46. Charles Hall and Kent Klitgaard, *Energy and the Wealth of Nations: An Introduction to Biophysical Economics*, 2nd ed. (New York: Springer, 2018), 213.

Chapter Seven

1. See Robert Heilbronner, *The Worldly Philosophers* (New York: Simon and Schuster, 1999).

2. These outer reaches can be identified from the physiocrats to John Maynard Keynes. Physiocrats identified surplus originating in agriculture, Malthus the possibility of limits on food production to keep pace with population, Marx a metabolic rift, and Arthur Cecil Pigou the presence of externalities. William Stanley Jevons gave us Jevon's paradox. Veblen identified emulative and conspicuous consumption, and Paul Baran and Paul Sweezy the necessity of waste to avoid stagnation. Keynes envisioned an end to expansion. And John Stuart Mill waxed poetic about the time when economic expansion would be halted to retain the magic of the Earth: he thought maybe it would be better to stop growth sooner as opposed to later. And no economist ever doubted that shortages of whatever kind would be reflected in prices.

3. Heilbronner also covered Joseph Schumpeter in his book. Schumpeter is not covered here.

4. Bill McKibben, *Falter* (New York: Henry Holt, 2019), 236.

5. See Adam Smith, *An Inquiry into the Nature and Causes of the Wealth of Nations* (Chicago: University of Chicago Press, 1976), 7, 17.

An incisive critique of Smith on this account comes from Eric Roll who tells us that Smith understands that the division of labor makes humans more productive but confuses the cause and effect of the division of labor because "it is not true, at least in theory, that division of labor requires the existence of private exchange." Therefore, "Adam Smith was guilty of making the characteristics of the society of his own day valid for all time; he regarded as a natural human motive and made into a universal principle of explanation, a feature of contemporaneous social order which was historically conditioned." See Eric Roll, *A History of Economic Thought*, 3rd ed. (Englewood Cliffs, NJ: Prentice-Hall, 1956), 154–55.

6. When Malthus and Ricardo wrote, the price of food was directly linked to the wage rate simply because the working class lived at a subsistence level. If the price of food went up, wages must go up: otherwise the working class would starve. When the Napoleonic war abated there existed a possibility to once again import grain (corn) from the continent, thereby opening up this discussion of the benefits of trade (and whether the Corn Laws should be repealed).

7. Thomas Robert Malthus, *An Essay on the Principle of Population* (New York: Dutton, 1961); David Ricardo, *Principles of Political Economy and Taxation* (London: Dent, 1962).

8. See Karl Marx, *Capital: A Critique of Political Economy*, ed. Frederick Engels (New York: Random House, 1906); Karl Marx, *Economic and Philosophic Manuscripts of 1844*, ed. Dirk J. Struik (New York: International, 1973); Karl Marx, "The German Ideology," in *The Marx Engels Reader*, ed. Robert C. Tucker (New York: W. W. Norton, 1972), 146–202; Friedrich Engels, "On the Division of Labor in Production," in *The Marx Engels Reader*, ed. Robert C. Tucker (New York: W. W. Norton, 1972), 718–24. Marx's analysis was extended by Paul Baran and Paul Sweezy into the monopoly stage of capitalism. See Paul Baran and Paul Sweezy, *Monopoly Capital* (New York: Monthly Review, 1966). Later Marxist scholars extended his work into the ecological realm. More will be said on this later.

9. William Stanley Jevons, *The Theory of Political Economy*, 1st ed. (London: Macmillan, 1871); Leon Walras, *Elements of Pure Economics* (Homewood, IL: Irwin, 1957); Carl Menger, *Principles of Economics* (New York: Free Press, 1950).

10. Thorstein Veblen, *Essays in Our Changing Order* (New York: Augustus M. Kelley, 1964).

11. Thorstein Veblen, *Engineers and Price System* (New York: Viking, 1933).

12. Thorstein Veblen, *The Theory of the Leisure Class* (New York: Dover, 1994).

13. John Maynard Keynes, *The General Theory of Employment, Interest and Money* (New York: Harcourt Brace, 1936).

14. Reducing taxes, deregulation, hyper-globalization, and increased privatization are the backbone of neoliberal economics.

15. Salination of soil, siltation of irrigation systems, deforestation, soil erosion come to mind.

16. This is another Wes Jackson aphorism.

17. It is important to state that EE is a broad category. There are many people in EE who are critical of it and yet consider themselves part of the ecological economic movement. Having said that, most ecological economists would say their theoretical core goes back to Herman Daly and Nicholaus Georgescu-Roegen. See Herman Daly, *Steady State Economics: The Economics of Biophysical Equilibrium and Moral Growth* (San Francisco: W. H. Freeman, 1977); Nicholas Georgescu-Roegen, *The Entropy Law and the Economic Process* (Cambridge, MA: Harvard University Press, 1976).

18. This orientation might be explained by G-R's and Daly's premature rejection of Marx and their aversion to socialism.

19. Nicholas Georgescu-Roegen, "Energy and Economic Myths" *Southern Economic Journal* 41, no. 3 (1975), 353.

20. Georgescu-Roegen, "Energy and Economic Myths," 348.

21. Georgescu-Roegen, "Energy and Economic Myths," 369.

22. Georgescu-Roegen, "Energy and Economic Myths," 363.

23. See, for example, Herman E. Daly, "Economics in a Full World," *Scientific American*, September 2005, 100–107.

24. Daly also says we should limit the money supply by keeping fractional reserves that banks must keep on deposits to 100 percent.

25. Herman Daly, "Reply to Troy Vettese's 'Against Steady-State Economics,'" *Ecological Citizen* 4 (2020), 80.

26. There are many critiques of Daly's work. See, for example: Richard Smith, "Beyond Growth or Beyond Capitalism?," *Real-World Economic Review*, no. 53 (2010): 28–42; Lisi Krall and Kent Klitgaard, "Ecological Economics and Institutional Change," *Annals of the New York Academy of Sciences* 1219 (2011): 185–96; Kent Klitgaard and Lisi Krall, "Ecological Economics, Degrowth, and Institutional Change," *Ecological Economics* 84(C) (2011): 247–53; John Bellamy Foster and Paul Burkett, *Marx and the Earth: An Anti-Critique* (Chicago, IL: Haymarket Books, 2017). Troy Vettese, "Against Steady-State Economics," *Ecological Citizen* 3 (Suppl. B) (2020): 35–46.

27. Georgescu-Roegen did not endorse the shift toward valuation and understood it for what it was—a palliative that would be ineffective in the end. The very categories of natural capital and ecosystem services are problematic. For example, one would assume that a term like "capital" is clearly understood and its meaning agreed upon, but nothing could be farther from the truth. In fact, some of the greatest controversies in the history of economic thought revolve around the meaning of capital. Here are some examples of the various definitions of capital: Daly defines capital as "a stock that yields a flow." Heilbronner describes it as "not a material thing but a process that uses material things as moments in its continuously dynamic existence." And Veblen tells us, "It is plain that, if the concept of capital were elaborated from observation of current business practice, it would be found that 'capital' is a pecuniary fact, not a mechanical one . . . and that the specific marks of capital, by which it is distinguishable from other facts, are of an immaterial character." There is clearly a lack of consensus as to the meaning of capital. And now, of course, we seem to be stuck with one definition in the lexicon of EE—a stock that yields a flow—likely the most unenlightening of the definitional possibilities if understanding duality and the emergence of a system at war with Earth is important.

28. David Attenborough, "Forward" in *The Economics of Biodiversity: The Dasgupta Review*, ed. Partha Dasgupta, 1–2. (London: HM Treasury, 2021).

29. For one discussion on this difficulty, See Robert W. Jensen, "Who is *We?*" *Ecological Citizen* 4 (2020): 57–61.

30. The list of scholars working in the area is extensive and includes James O'Connor, Joel Kovel, Gar Alperovits, John Bellamy Foster, Jason W. Moore, Richard Smith, Clive Spash, Giorgos Kallis, Kent Klitgaard, Lisi Krall, and many more.

31. Jason Moore, "The Rise of Cheap Nature," in *Anthropocene or Capitalocene? Nature, History, and the Crisis of Capitalism*, ed. Jason Moore (Oakland, CA: PM Press, 2016), 78–115, 79.

32. Jason Moore, "The Capitalocene Part I: On the Nature and Origins of Our Ecological Crisis," *Journal of Peasant Studies* 44, no. 3 (2017): 594–630.

33. Moore claims capitalism turned most humans and nature into the same category, all "regarded as part of Nature, along with trees and soils and rivers—and treated accordingly." The same processes that produce class and inequality also produce ecological degradation, and this is all part of the environment-making process of capitalism—it is part of capitalism's "world ecology." See Moore, "Cheap Nature," 79.

34. Jason Moore and Raj Patel, *A History of the World in Seven Cheap Things* (Oakland: University of California Press, 2017). The charge that humans had a hand in in the large-scale extinction of Pleistocene mammals is a rather complicated story and anything but an accepted verity of history. And the implicit assumption that humans have a natural propensity to overshoot is similarly controversial. The truth is that humans are a contextual species, and the context of hunting and gathering did not predispose humans to overshoot.

35. Paul Shepard, "Ten Thousand Years of Crisis," in *The Only World We've Got*, ed. Paul Shepard (San Francisco: Sierra Club, 1996), 190.

36. Moore, "Cheap Nature," 114.

37. More than two decades ago James O'Connor expanded Marx's idea of the contradiction of capitalism by introducing the second contradiction of capitalism. See James O'Connor, *Natural Causes: Essays in Ecological Marxism* (New York: Guilford, 1998). The first contradiction is that the accumulation of capital undermines the ability for further accumulation of capital—in simplistic terms the necessity to hold wages down and replace workers with machines would undermine the ability to realize profit. The accumulation process would be undermined by the accumulation process in a distinct way with regard to the second contradiction. Here accumulation would be undermined by the destabilizing effect of ecological degradation; yet ecological degradation was part and parcel of the accumulation process where competition for market share would force producers to ignore ecological limits. This is pretty straightforward and simple to understand and is connected to the idea that Foster and others highlight—that of capitalism's metabolic rift.

38. Foster and Burkett, *Marx and the Earth*, 70.

39. Wes Jackson tells us: "So destructive has the agricultural revolution been that, geologically speaking, it surely stands as the most significant and explosive event to appear on the face of the earth, changing the earth even faster than did the origin of life." See Wes Jackson, *Nature as Measure* (Berkeley, CA: Counterpoint, 2011), 5.

40. Engels was clearly onto a development of what later became gene-culture coevolution as Bellamy Foster points out, but at the end of the eighteenth century that conversation was more constrained by knowledge and ideas than it is today.

41. Paul Shepard, *Nature and Madness* (San Francisco: Sierra Club, 2006), 23.

42. Foster and Burkett, *Marx and the Earth*, 6–7. For Marx, one obvious indication of the separation of humans from their relationship to Earth was found

in the concentration of humans in urban centers separate from the country. Wastes were not recycled, and soil was degraded.

43. Foster and Burkett, *Marx and the Earth*, 239.

44. Roll, *History of Economic Thought*, 14.

45. In their paper, "Bridging Ecological and Social Systems Coevolution," Gual and Norgaard call for a new method in economic thinking—one that acknowledges the existence of "interlinked/interdependent evolutionary processes between cultural and biotic systems." Miguel A. Gual and Richard B. Norgaard, "Bridging Ecological and Social Systems Coevolution: A Review and Proposal," *Ecological Economics* 69 (2010): 707–17. John Gowdy and I have done extensive work on human ultrasociality.

Epilogue

1. Paul Raskin, "Interrogating the Anthropocene: Truth or Fallacy," opening essay for a GTI Forum, *Great Transition Initiative* (February 2021).

2. Lisi Krall, "Resistance," in *Keeping the Wild: Against the Domestication of Earth,* ed. George Wuerthner, Eileen Crist, and Tom Butler (Washington, DC: Island, 2014), 205–210.

3. William Cronon, "The Trouble With Wilderness; or, Getting Back to the Wrong Nature," in *Uncommon Ground: Rethinking the Human Place in Nature*, ed. William Cronon(New York: W. W. Norton, 1995), 80, 83.

4. Peter Kareiva, Michelle Marvier, and Robert Lalasz, "Conservation in the Anthropocene: Beyond Solitude and Fragility," *Breakthrough Journal* no. 2 (Winter 2012), 8, 2–3.

These idealized notions are seen as deeply flawed according to Kareiva and others: "The wilderness so beloved by conservationists—places 'untrammeled by man'—never existed, at least not in the last thousand years." Kareiva, "Conservation," 5. I'm not sure where Kareiva gets the data to support this claim. In every century since the advent of agriculture there has been a diminishment of untrammeled wilderness, but a thousand years ago when there were about five hundred million people on the planet (compared with almost eight billion now), a lot of the earth was "untrammeled by man."

5. Kareiva et al., "Conservation," 2.

6. Rebecca Solnit, "John Muir in Native America," *Sierra Club Magazine*, March–April 2021, 2–3.

7. Solnit, "John Muir," 2–3.

8. Jefferson quotation found in Lisi Krall, *Proving Up: Domesticating Land in US History* (Albany: State University of New York Press, 2010), 25.

9. Terminology attributed to Nicholas Georgescu-Roegen, "Energy and Economic Myths," *Southern Economic Journal* 41, no. 3 (1975): 347–81.

10. Solnit, "John Muir," 6.

11. Even so, in the Tetons the Yellowstone country that runs adjacent humans are not at the top of the food chain.

12. The second law of thermodynamics is the Entropy Law, where entropy is "a measure of the unavailable energy in closed thermodynamic system." See Nicholas Georgescu-Roegen, *The Entropy Law and the Economic Process* (Cambridge, MA: Harvard University Press. 1971), 5.

13. Kent Klitgaard, "Hydrocarbons and the Illusion of Sustainability," *Monthly Review* 68, no. 3 (2016): 1–17.

14. More recently Keynesian economics and the financing of the Green New Deal have been bolstered by the rise of Modern Monetary Theory (MMT). MMT would argue that it isn't necessary to bother with that politically charged bugaboo of deficit spending. The government can essentially spend as much money as it wants because the economic system has a lot of slack before inflationary pressures set in and because the government essentially has the power to create the money to do so.

15. It is important to keep in mind the limits of technology. Georgescu-Roegen pointed out that "economic history confirms a rather elementary fact—the fact that the great strides in technological progress have generally been touched off by a discovery of how to use a new kind of accessible energy. On the other hand, a great stride in technological progress cannot materialize unless the corresponding innovation is followed by a great mineralogical expansion." See Nicholas Georgescu-Roegen, "Energy and Economic Myths," in *Southern Economic Journal* 41, no. 3 (1975): 361.

16. For perhaps the best discussion of the limitations and confusion of the Green New Deal see Stan Cox, *The Green New Deal and Beyond* (San Francisco: City Lights, 2020).

17. Georgescu-Roegen, "Energy and Economic Myths," 371.

18. The presentation of global warming as a technological problem with a technological solution hides just about everything we need to acknowledge. The prospect of 100 percent renewable energy by 2050 has gotten a lot of press. It depends on very high return to energy efficiency (lowering demand for energy while the economy grows), and it completely ignores present distributional problems in access to energy.

See Mark Z. Jacobson and Mark A. Delucchi, "Providing All Global Energy with Wind, Water, and Solar Power: Part I, Technologies, Energy Resources, Quantities and Areas of Infrastructure, and Materials," in *Energy Policy* 39 (2011): 1154–69; Mark Z. Jacobson and Mark A. Delucchi, "Providing All Global Energy with Wind, Water, and Solar Power: Part II, Reliability, System, and Transmission Costs, and Policies," *Energy Policy* 39 (2011): 1170–90. There is much information emerging about the impossibility of this technological feat. In addition to Cox see Ted Trainer, "A Critique of Jacobson and Delucchi's Proposals for a World Renewable Energy Supply," *Energy Policy* 44 (2012): 476–81. See also Alexander E. Macdonald et al.,

"Future Cost-competitive Electricity Systems and Their Impact on US CO_2 Emissions," *Nature Climate Change* 6, no. 4 (January 2016): 526–31. See also Shoibal Chakravarty et al., "Sharing Global CO_2 Emission Reductions Among One Billion High Emitters," *PNAS* 106, no. 29 (2009): 11884–888.

19. Pinker, *Enlightenment Now*, 23.

Bibliography

Abu-Lughod, Janet. "The Shape of the World System in the Thirteenth Century." *Studies in Comparative International Development* 22, no. 4 (1988): 3–24.

Arrighi, Giovanni. *Adam Smith in Beijing: Lineages of the Twenty-First Century.* London: Verso, 2007.

Attenborough, David. "Forward." In *The Economics of Biodiversity: The Dasgupta Review.* Abridged version. edited by Partha Dasgupta, 1–2. London: HM Treasury, 2021.

Bar-Yosef, Ofer. "The Natufian Culture in the Levant, Threshold to the Origins of Agriculture." *Evolutionary Anthropology* (December 1998): 159–77.

Baran, Paul, and Paul Sweezy. *Monopoly Capital.* New York: Monthly Review Press, 1966.

Beerling, David J. "New Estimates of Carbon Transfer to Terrestrial Ecosystems Between the Last Glacial Maximum and the Holocene." *Terra Nova* 11 (1999): 162–67.

Berry, Wendell. "The Peace of Wild Things." In *The Peace of Wild Things: And Other Poems.* London: Penguin, 2018.

Beshers, Samuel N., and Jennifer H. Fewell. "Models of Division of Labor in Social Insects." *Annual Review of Entomology* 46 (2001): 413–40.

Boehm, Christopher. *Moral Origins: The Evolution of Virtue, Altruism, and Shame.* New York: Basic Books, 2011.

Boomsma, Jacobus, J., and Nigel R. Franks. "Social Insects: From Selfish Genes to Self-Organization and Beyond." *Trends in Ecology and Evolution* 21, no. 6 (2006): 303–308.

Bowles, Samuel. "Warriors, Levelers, and the Role of Conflict in Human Social Evolution." *Science* 336 (2012): 876–79.

Bowles, Samuel, and Jung-Kyoo Choi. "Coevolution of Farming and Private Property during the Early Holocene." *Proceedings of the National Academy of Sciences USA* 110, no. 22 (2013): 8830–35.

Bowles, Samuel, and Herbert Gintis. *A Cooperative Species: Human Reciprocity and its Evolution.* Princeton, NJ: Princeton University Press, 2011.

Braverman, Harry. *Labor and Monopoly Capital.* New York: Monthly Review, 1974.

Campbell, Donald T. "Legal and Primary-Group Social Controls." In *Law, Biology and Culture: The Evolution of Law*, edited by Margaret Gruter and Paul Bohannan, 59–71. Berkeley, CA: Bepress, 1982.

Chakravarty, Shoibal, Ananth Chikkatur, Heleen de Coninck, Stephen Pacala, Robert Socolow, and Massimo Tavoni. "Sharing Global CO_2 Emission Reductions Among One Billion High Emitters." *PNAS* 106, no. 29 (2009): 11884–888.

Cohen, Mark, and Gillian Crane-Kramer. *Ancient Health: Skeletal Indicators of Agricultural and Economic Intensificiation.* Gainesville: University of Florida Press, 2007.

Corning, Peter. "Holistic Darwinism: 'Synergistic Selection' and the Evolutionary Process." *Journal of Social and Evolutionary Systems* 20, no. 4 (1997): 363–400.

———. *Synergistic Selection: How Cooperation Has Shaped Evolution and the Rise of Humankind.* Hackensack, NJ: World Scientific, 2018.

Cox, Thomas S. "Crop Domestication and the First Plant Breeders." In *Plant Breeding and Farmer Participation*, edited by Salvatore Ceccarelli, Elcio Guimaraes, and Eva Weltzien, 1–26. Rome: Food and Agricultural Organization, 2009.

Cox, Stan. *The Green New Deal and Beyond.* San Francisco: City Lights, 2020.

Crist, Eileen. *Abundant Earth: Toward an Ecological Civilization.* Chicago: University of Chicago Press, 2019.

Cronon, William. "The Trouble with Wilderness; or, Getting Back to the Wrong Nature." In *Uncommon Ground: Rethinking the Human Place in Nature.* edited by William Cronon, 69–90. New York: W. W. Norton, 1995.

Crosby, Alfred W. Jr. *The Colombian Exchange.* Westport, CT: Greenwood, 1972.

Daly, Herman. "Reply to Troy Vettese's 'Against Steady-state Economics,'" *Ecological Citizen* 4 (2020): 79–82.

———. "Economics in a Full World," *Scientific American* (September 2005): 100–107.

———. *Steady State Economics: The Economics of Biophysical Equilibrium and Moral Growth.* San Francisco: W. H. Freeman, 1977.

Diamond, Jared. "The Worst Mistake in the History of the Human Race." *Discover*, May 1987, 64–66.

Diamond, Stanley. *In Search of the Primitive: A Critique of Civilization.* New Brunswick, NJ, and London: Transaction, 1974.

Duarte, Ana, Franz J. Weissing, Ido Pen, and Laurent Keller. "An Evolutionary Perspective on Self-Organized Division of Labor in Social Insects," *Annual Review of Ecology, Evolution, and Systematics* 42 (2011): 91–110.

Durkheim, Emile. *The Division of Labor in Society.* New York: Free Press, 1964.

Edward, Richard. *Contested Terrain: The Transformation of the Workplace in the Twentieth Century.* New York: Basic Books, 1979.

Engels, Friedrich. "On the Division of Labour in Production." In the *Marx-Engels Reader*, edited by Robert C. Tucker, 718–24. New York: W. W. Norton, 1972.

Fewell, Jennifer H. "Social Insect Networks." *Science* 301 (2003): 1867–70.

Flannery, Tim. "The Superior Civilization." *New York Review of Books*, February 26, 2009.

Foster, John Bellamy, and Paul Burkett. *Marx and the Earth: An Anti-Critique.* Chicago: Haymarket, 2017.

Frank, Andre Gunder. "A Theoretical Introduction to 5000 Years of World System History." *Review* 13 (1990): 155–248.

Georgescu-Roegen, Nicholas. "Energy and Economic Myths," *Southern Economic Journal* 41, no. 3 (1975): 347–80.

———. *The Entropy Law and the Economic Process.* Cambridge, MA: Harvard University Press, 1971.

Gills, Barry, and Andre Frank. "5000 Years of World System History: The Cumulation of Accumulation." In *Core/Periphery Relations in Precapitalist Worlds*, edited by Christopher Chase-Dunn and Thomas D. Hall, 67–112. Boulder, CO: Westview, 1991.

Gintis, Herbert. *Individuality and Entanglement: The Moral and Material Bases of Social Life.* Princeton, NJ: Princeton University Press, 2017.

Gordon, Deborah M. "Control Without Hierarchy." *Nature* 446, 8 (2007): 143.

———. "From Division of Labor to the Collective Behavior of Social Insects." *Behavioral Ecology and Sociobiology* 70 (2016): 1101–1108.

Gould, Stephen J. *Ever Since Darwin.* New York: W. W. Norton, 1973.

Gowdy, John, and Lisi Krall. "Agriculture as a Major Evolution Transition to Human Ultrasociality." *Journal of Bioeconomics* 16 (2014): 179–202.

———. "Disengaging from the Ultrasocial Economy: The Challenge of Directing Evolutionary Change." *Behavioral and Brain Sciences* 39 (2016): 41–62.

———. "The Economic Origins of Ultrasociality." *Behavioral and Brain Sciences* 39 (2016): 1–41.

———. "The Ultrasocial Origin of the Anthropocene." *Ecological Economics* 95 (2013): 137–47.

Graeber, D. *Debt: The First 5000 Years.* Brooklyn, NY: Melville House, 2012.

Grimes, P. "Evolution and World-Systems: Complexity, Energy, and Form." *Journal of World-Systems Research* 23, no. 2 (2017): 678–732.

Grinsted, Lena, Ingi Agnarsson, and Trine Bilde. "Subsocial Behaviour and Brood Adoption in Mixed-Species Colonies of Two Theridiid Spiders." *Naturwissenschaften* 99 (2012): 1021–30.

Gual, Miguel A., and Richard B. Norgaard. "Bridging Ecological and Social Systems Coevolution: A Review and Proposal." *Ecological Economics* 69 (2010): 707–717.

Hall, Charles A. S. "Energy Return on Investment." In *The Energy Reader: Overdevelopment and the Delusion of Endless Growth.* edited by Tom Butler, Daniel Lerch, and George Wuerthner, 62–68. Sausalito, CA: Watershed Media, 2012.

Hall, Charles A. S., and Kent A. Klitgaard. *Energy and the Wealth of Nations: An Introduction to Biophysical Economics.* 2nd ed. New York: Springer, 2018.

Harari, Yuval N. *Sapiens: A Brief History of Humankind.* London: Vintage, 2011.

Heilbronner, Robert. *The Worldly Philosophers*. New York: Simon and Schuster, 1999.

Henrich, Joseph. "Cultural Group Selection, Coevolutionary Processes and Large-Scale Cooperation." *Journal of Economic Behavior and Organization* 53 (2004): 3–35.

Hobsbawm, Eric. *The Age of Revolution 1789–1848*. London: Weidenfeld and Nicolson, 1962.

Holbrook, C. Tate, Phillip M. Barden, and Jennifer H. Fewell. "Division of Labor Increases with Colony Size in Harvester Ant *Pogonmyrmex californicus*." *Behavioral Ecology* 22, no. 5 (2011): 960–66.

Hölldobler, Bert, and Edward O. Wilson. *The Leafcutter Ants: Civilization by Instinct*. New York: W. W. Norton, 2011.

———. *The Superorganism*. New York: W. W. Norton, 2009.

Hunt, Emery K. *Property and Prophets: The Evolution of Economic Institutions and Ideologies*. Updated 7th Edition. Armonk, NY: M. E. Sharpe. 2003.

Jablonka, Eva, and Marion Lamb. *Evolution in Four Dimensions*. Cambridge, MA: MIT Press, 2014.

Jackson, Wes. *Consulting the Genius of the Place*. Berkeley: Counterpoint, 2010.

———. "Five Carbon Pools." In *The Energy Reader: Overdevelopment and the Delusion of Endless Growth*, edited by Tom Butler, Daniel Lerch, and George Wuerthner, 27–32. Healdsburg, CA: Watershed Media, 2012.

———. *Nature as Measure*. Berkeley: Counterpoint, 2011.

Jacobson, Mark Z., and Mark A. Delucchi. "Providing All Global Energy with Wind, Water, and Solar Power: Part I, Technologies, Energy Resources, Quantities and Areas of Infrastructure, and Materials." *Energy Policy* 39 (2011): 1154–69.

———. "Providing All Global Energy with Wind, Water, and Solar Power: Part II, Reliability, System, Transmission Costs, and Policies." *Energy Policy* 39 (2011): 1170–90.

Jeanne, Robert L. "Division of Labor is Not a Process or a Misleading Concept." *Behavioral Ecology and Sociobiology* 70, no. 7 (2016): 1109–12.

Jensen, Robert. "Who is *We*?" *The Ecological Citizen* 4 (2020): 57–61.

Jevons, William Stanley. *The Theory of Political Economy*. 1st ed. London: Macmillan, 1871.

Kareiva, Peter, Michelle Marvier, and Robert Lalasz. "Conservation in the Anthropocene." *Breakthrough Journal* 2 (Winter 2012): 1–14.

Keller, Laurent. *Levels of Selection in Evolution*. Princeton, NJ: Princeton University Press, 1999.

Keynes, John Maynard. *The General Theory of Employment, Interest and Money*. New York: Harcourt Brace, 1936.

Klitgaard, Kent. "Hydrocarbons and the Illusion of Sustainability." *Monthly Review* 68, no. 3 (2016): 77–88.

Klitgaard, Kent, and Lisi Krall. "Ecological Economics, Degrowth, and Institutional Change." *Ecological Economics,* 84(C) (2011): 247–53.

Krall, Lisi. "The Economic Legacy of the Holocene." *Ecological Citizen* 2 (2018): 67–76.

———. *Proving Up: Domesticating Land in US History*. Albany: State University of New York Press, 2010.

———. "Reckoning our Ultrasocial Past." *Trumpeter* 31, no. 2 (2015): 102–111.

———. "Resistance." In *Keeping the Wild: Against the Domestication of Earth*, edited by George Wuerthner, Eileen Crist, and Tom Butler. 205–210. Washington, DC: Island, 2014.

Krall, Lisi, and Kent Klitgaard. "Ecological Economics and Institutional Change." Annals of New York Academy of Sciences 1219 (2011): 185–96.

Laland, Kevin. *Darwin's Unfinished Symphony: How Culture Made the Human Mind*. Princeton, NJ: Princeton University Press, 2017.

———. "Evolution Unleashed: Is Evolutionary Science Due for a Major Overhaul—Or Is Talk of 'Revolution' Misguided?" *Aeon*, 2018, 1–12.

Lambert, Patricia M. "Health Versus Fitness." *Current Anthropology*, 50, no. 5 (2009): 603–608.

Landes, David. *The Unbound Prometheus: Technological Change and Industrial Development in Western Europe from 1750 to the Present*. Cambridge: Cambridge University Press, 1969.

Larsen, Clark Spencer. "The Agricultural Revolution as Environmental Catastrophe: Implications for Health and Lifestyles in the Holocene." *Quarterly International* 150 (2006): 12–20.

Lee, Richard. "Non-Capitalist Work: Baseline for an Anthropology of Work or Romantic Delusion?" *Anthropology of Work Review* 18, no. 4 (1998): 9–13.

Levi-Strauss, Claude. *The Savage Mind*. Chicago: University of Chicago Press, 1962.

Lewis, Simon, and Mark Maslin. *The Human Planet: How We Created the Anthropocene*. New Haven, CT: Yale University Press, 2018.

Macdonald, Alexander E., Christopher T. M. Clack, Anneliese Alexander, Adam Dunbar, James Wilczak, and Yuanfu Xie. "Future Cost-Competitive Electricity Systems and Their Impact on US CO_2 Emissions." *Nature Climate Change* 6 (January 2016): http://doi.org/10.1038/nclimate2921.

Malm, Andreas. *Fossil Capital: The Rise of Steam Power and the Roots of Global Warming*. London: Verso, 2016.

Malthus, Thomas Robert. *An Essay on the Principle of Population*. New York: Dutton, 1961.

Margulis, Lynn. *Origin of Eukaryotic Cells*. New Haven, CT: Yale University Press, 1970.

Marx, Karl. *Capital: A Critique of Political Economy, Vol. 1: The Process of Capitalist Production*. New York: International Publishers, 1979.

———. *Economic and Philosophic Manuscripts of 1844*. Edited by Dirk J. Struik. New York: International Publishers, 1973.

———. "The German Ideology." In *The Marx-Engels Reader*, edited by Robert C. Tucker, 146–202. New York: W. W. Norton, 1972.

McCorriston, Joy, and Frank Hole. "The Ecology of Seasonal Stress and the Origins of Agriculture in the Near East." *American Anthropologist, New Series* 93, 1 (1991): 46–69.

McKibben, Bill. *Falter.* New York: Henry Holt, 2019.

Mead, George Herbert. *Mind, Self and Society: The Definitive Edition.* Chicago: University of Chicago Press, 2015.

Menger, Carl. *Principles of Economics.* New York: Free Press, 1950.

Midgley, Mary. *Beast and Man: The Roots of Human Nature.* Abingdon, UK: Routledge, 2002.

Moffett, Mark W. *Adventures Among Ants: A Global Safari with a Cast of Trillions.* Berkeley: University of California Press, 2010.

———. "Human Identity and the Evolution of Societies." *Human Nature* 24 (2013): 219–67.

Moore, Jason W. "Ecology, Capital, and the Nature of Our Times: Accumulation and Crisis in the Capitalist World-Ecology." *American Sociology Association,* 28, no. 1 (2011): 108–47.

———. "The Capitalocene Part I: On the Nature and Origins of Our Ecological Crisis." *Journal of Peasant Studies,* April (2017): 1–37.

———. "The Rise of Cheap Nature." In *Anthropocene or Capitalocene? Nature, History, and the Crisis of Capitalism,* edited by Jason Moore, 78–115. Oakland, CA: PM Press/Kairos, 2016.

Moore, Jason W., and Raj Patel. *A History of the World in Seven Cheap Things.* Oakland: University of California Press, 2017.

Mueller, Ulrich G., and Christian Rabeling. "A Breakthrough Innovation in Animal Evolution," *PNAS* 105, no. 14 (2008): 5287–88.

North, Douglass C., and Paul Thomas Robert. "The First Economic Revolution." *Economic History Review,* 30, no. 2 (1977): 229–41.

Nowak, Martin, Corina Tarnita, and Edward O. Wilson. "The Evolution of Eusociality." *Nature* 466 (2010): 1057–62.

O'Connor, James. *Natural Causes: Essays in Ecological Marxism.* New York: Guilford, 1998.

Okasha, Samir. *Evolution and the Levels of Selection.* Oxford: Clarendon, 2006.

Pinker, Steven. *Enlightenment Now: The Case for Reason, Science, Humanism, and Progress.* New York: Penguin, 2018.

Pigliucci, Massimo, and Gerd B. Müller, eds. *Evolution: The Extended Synthesis.* Cambridge, MA: MIT Press, 2010.

Polanyi, Karl. *The Great Transformation.* Boston: Beacon, 1957.

Raskin, Paul. "Interrogating the Anthropocene: Truth or Fallacy." GTI Forum, *Great Transition Initiative,* February 2021.

Ricardo, David. *Principles of Political Economy and Taxation.* London: Dent, 1962.

Richerson, Peter, Robert Boyd, and Robert Bettinger. "Was Agriculture Impossible during the Pleistocene but Mandatory during the Holocene? A Climate Change Hypothesis." *American Antiquity* 66 (2001): 387–411.

Richerson, Peter, and Robert Boyd. *Not by Genes Alone: How Culture Transformed Human Evolution*. Chicago: University of Chicago Press, 2005.

Robson, Simon K. A., and James F. A. Traniello. "Division of Labor in Complex Societies: A New Age of Conceptual Expansion and Integrative Analysis." *Behavioral Ecology and Sociobiology* 70 (2016): 995–98.

Roll, Eric. *A History of Economic Thought*. 3rd ed. New Brunswick, NJ: Prentice-Hall, 1956. 154–55.

Safina, Carl. *Beyond Words: What Animals Think and Feel*. New York: Henry Holt, 2015.

Sahlins, Marshall. "The Original Affluent Society." In *Limited Wants Unlimited Means: A Reader on Hunter-Gatherer Economics and the Environment*. edited by John Gowdy. Washington, DC: Island, 1998.

———. *Stone Age Economics*. London: Routledge, 1974.

Sale, Kirkpatrick. *After Eden: The Evolution of Human Domination*. Durham, NC: Duke University Press, 2006.

Scott, James. *Against the Grain: A Deep History of the Earliest States*. New Haven, CT: Yale University Press, 2017.

Schumpeter, Joseph. *History of Economic Analysis*. New York: Oxford University Press, 1986.

Sears, Paul. *Deserts on the March*. Norman: University of Oklahoma Press, 1959.

Schultz, Ted, and Sean G. Brady. "Duarte, Major Evolutionary Transition in Ant Agriculture." *Proceedings of the National Academy of Sciences* 105 (2008).

Schultz, Ted, Ulrich G. Mueller, Cameron R. Currie, and Stephen A. Rehner. "Reciprocal Illumination: A Comparison of Agriculture in Humans and in Fungus-Growing Ants." In *Ecological and Evolutionary Advances in Insects-Fungal Associations*, edited by Fernando Vega and Meredith Blackwell, 149–90. New York: Oxford University Press, 2005.

Shepard, Paul. *Coming Home to the Pleistocene*. edited by Florence R. Shepard. Washington DC: Island/Shearwater, 1998.

———. *Nature and Madness*. San Francisco: Sierra Club, 1982.

———. "Ten Thousand Years of Crisis." In *The Only World We've Got*, edited by Paul Shepard, 190–211. San Francisco: Sierra Club, 1996.

———. *The Tender Carnivore and the Sacred Game*. New York: Scribner's, 1973.

Smith, Adam. *An Inquiry into the Nature and Causes of the Wealth of Nations*. Chicago: University of Chicago Press, 1976.

Smith, John, and Eörs Szathmáry. *The Origins of Life: From the Birth of Life to the Origins of Language*. Oxford: Oxford University Press, 1999.

Smith, Richard. "Beyond Growth or Beyond Capitalism?" *Real-World Economic Review* 53 (2010): 28–42.

Solnit, Rebecca. "John Muir in Native America." *Sierra Club Magazine*, March–April 2021.

Streit-Krug, Aubrey. "Ecospheric Care Work." *Ecological Citizen* 3, no. 2 (2020): 143–48.

Tainter, Joseph A. "Sustainability of Complex Societies." *Futures* 27, no. 4 (1995): 397–407.

Thompson, Edward. *The Making of the English Working Class*. New York: Vintage, 1963.

Trainer, Ted. "A Critique of Jacobson and Delucchi's Proposals for a World Renewable Energy Supply." *Energy Policy* 44 (2012): 476–81.

Veblen, Thorstein. *Absentee Ownership and the Business Enterprise in Recent Times*. New York: Sentry, 1964.

———. *Essays in Our Changing Order*. New York: Augustus Kelley, 1964.

———. *The Theory of the Leisure Class*. New York: Dover, 1994.

Vettese, Troy. "Against Steady-State Economics." *Ecological Citizen* 3 (Suppl. B) (2020): 35–46.

Vitek, Bill. "Dandelions are Divine." *Ecological Citizen* 2 (2019): 189–93.

———. "God or Nature: Desire and the Quest for Unity." *Minding Nature* 4, 2 (2011): 20–25.

Walras, Leon. *Elements of Pure Economics*. Homewood IL: Irwin, 1957.

Wexler, Bruce E. *Brain and Culture*. Cambridge, MA: MIT Press, 2006.

White, Lynn Jr. *Medieval Technology and Social Change*. Oxford: Oxford University Press, 1964.

Wilson, David Sloan. "Human Cultures are Primarily Adaptive at the Group Level." The Evolution Institute, 1–19. https://evolution-institute.org/focus-article/david-sloan-wilson.

———. "Laying the Foundation for Economics." Open Peer Commentary to Gowdy & Krall: "The Economic Origins of Ultrasociality." *Behavioral and Brain Sciences* 39 (2016): 40–41.

———. "Multilevel Selection and Major Transitions." In *Evolution: The Extended Synthesis*. edited by Massimo Pigliucci and Gerd Müller, 81–91. Cambridge, MA: MIT Press, 2010.

Wilson, David Sloan, and John M. Gowdy. "Human Ultrasociality and the Invisible Hand: Foundational Developments in Evolutionary Science Alter a Foundational Concept in Economics." *Journal of Bioeconomics* 17, no. 2 (2015): 37–52.

Wilson, Edward O. *"One Giant Leap: How Insects Achieved Altruism and Colonial Life."* Bioscience 58, no. 1 (2008): 17–25.

———. *The Social Conquest of the Earth*. New York: Liveright, 2012.

Wilson, Edward O., and David Sloan Wilson. "Rethinking the Theoretical Foundation of Sociobiology." *Quarterly Review of Biology* 82, no. 4 (2007): 327–48.

Wood, Ellen Meiksins. *The Origin of Capitalism*. New York: Monthly Review, 1999.

Wright, Ronald A. *Short History of Progress*. Cambridge, MA: DaCapo, 2004.

Index

Note: Page numbers followed by "n" indicate numbered endnotes.

agricultural revolution: change process after, 82–84, 119–20; connection between ecological declension and, 123; defined, 15–16, 133n1 (Intro), 134n2; downward spiral / deskilling associated with, 37–38, 41–42, 54–55, 58–60, 75–76, 146n3; as expression of cooperation, 30–34; impact of, 39–40, 116, 158n39; lack of consensus on causes of, 49, 145n28; rise of civilization and, 54; soil carbon and, 56–57, 58, 144n3; as "the fall" / "fault line in human history" (Jackson), ix–xi, 7, 55

agriculture (grain agriculture), 53–64; agricultural systems as economic superorganisms, xi, 5, 20–23, 53; agriculture as "universal" and, 2, 3, 5–6, 16–18, 29–30, 36, 53, 82, 146n1; animal husbandry and, 147n17; change in collective configuration of humans, 44; characteristics of, 35; coevolution in, 43, 44, 57–60, 147n15, 148n20; complex tapestry of, 56–57, 61; cooperation and coordination in, 60; cultivation as collective enterprise, 17, 29–31, 60;

cultivation techniques and tasks, 47, 59–61, 148n20; culture and, 60, 61; defined, 9; deskilling associated with, 37–38, 41–42, 58–60, 75–76, 146n3; division of labor in, 35–36, 53–64; duality between humans and more-than-humans and, 2–3, 17, 26, 41–42, 54, 55, 76–78, 98, 116–17; efficiency in, 60; expansionary spiral in, 11, 22–23, 39–40, 43, 57–58, 63; external conditions leading to, 53, 56–57; family size and structure, 148n22; feudalism and, 83–86; grain types in, 44, 58, 148n20; Holocene warming and, 44–45, 53, 55–56, 134n4, 141n3, 149n5; *Homo sapiens agriculturii* and (*see Homo sapiens agriculturii*); human inclinations vs., 41–42; impact of, 39–40, 74–78; irrigation systems, 17, 47, 57, 58, 60, 77, 128, 129; "just-so-humancentric" story vs., 2, 4, 16–17, 25; legacy of, 74–78; modern economic life and, 2–3; plowshares, 17, 58, 60; problems of (*see* problems of grain agriculture); seed selection in, 44, 140–41n2; self-referential dynamic in, 6–8,